Biotechnology
Intelligence
Unit

Enzyme Mixtures and Complex Biosynthesis

Sanjoy K. Bhattacharya, M. Tech., Ph.D.
Bascom Palmer Eye Institute
University of Miami
Miami, Florida, U.S.A.

Landes Bioscience
Austin, Texas
U.S.A.

ENZYME MIXTURES AND COMPLEX BIOSYNTHESIS

Biotechnology Intelligence Unit

Landes Bioscience

Printed in the U.S.A.

Please address all inquiries to the Publisher:
Landes Bioscience, 1002 West Avenue, 2nd Floor, Austin, Texas 78701, U.S.A.
Phone: 512/ 637 6050; Fax: 512/ 637 6079
www.landesbioscience.com

ISBN: 978-1-58706-216-2

Library of Congress Cataloging-in-Publication Data

Enzyme mixtures and complex biosynthesis / [edited by] Sanjoy K. Bhattacharya.
p. ; cm. -- (Biotechnology intelligence unit)
Includes bibliographical references and index.
ISBN 978-1-58706-216-2
1. Multienzyme complexes. 2. Enzymes--Synthesis. 3. Biotechnology. I. Bhattacharya, Sanjoy K., Ph. D. II. Series: Biotechnology intelligence unit (Unnumbered)
[DNLM: 1. Multienzyme Complexes. 2. Biotechnology--methods. 3. Protein Biosynthesis. QU 135 E60347 2007]
QP601.E5218 2007
660.6'34--dc22

2007033862

About the Editor...

SANJOY K. BHATTACHARYA is Assistant Professor of Ophthalmology and Molecular Neuroscience at the Bascom Palmer Eye Institute, University of Miami, Miami, Florida, U.S.A. A graduate of Indian Institute of Technology-Delhi, specializing in biochemical engineering, Dr. Bhattacharya's research is focused on understanding proteins and enzymes. His research interest includes protein structure function, proteomics, translational regulation of proteins and application of proteins and enzymes for complex biosynthesis. He is research editor for Microbial Cell factories (a Biomed central journal) and has been a member of the editorial board of the *Journal of Biophysics, Molecular Biology and Biochemistry*. He also has been Editor-in-Chief of *Recent Research Developments in Biotechnology and Bioengineering*. Dr. Bhattacharya is also credited with co-founding two small biotech enterprises, ABRD Company LLC and Inspiracom Biotech which focus using enzymes for complex biosynthesis and acts as their scientific advisor. He is also a co-founder and former president of non-profit Malignancy Research Foundation. Dr. Bhattacharya is a member of several national and international societies such as New York Academy of Science, DNA Methylation Society, The Association of Research in Vision and Ophthalmology (ARVO), American Society for Biochemistry and Molecular Biology (ASBMB) and Society for Neuroscience (SfN).

Dedication

Dedicated to those who made excellent contributions towards the advancement of science and engineering and yet remained anonymous

CONTENTS

EDITOR

Sanjoy K. Bhattacharya
Bascom Palmer Eye Institute
University of Miami
Miami, Florida, U.S.A.
Email: sbhattacharya@med.miami.edu
Chapters 1, 9

CONTRIBUTORS

Mabel Algeciras
Bascom Palmer Eye Institute
University of Miami
Miami, Florida, U.S.A.
Email: malgeciras@med.miami.edu
Chapter 9

Çiğdem Babaarslan
Department of Chemical Engineering
Ankara University Engineering Faculty
Ankara, Turkey
Chapter 7

Emine Bayraktar
Department of Chemical Engineering
Ankara University Engineering Faculty
Ankara, Turkey
Chapter 7

Katalin Bélafi-Bakó
Research Institute of Chemical and Process Engineering
University of Veszprem
Egyetem, Hungary
Email: bako@delta.richem.hu
Chapter 4

Diana Braikova
Bulgarian Academy of Sciences
Institute of Organic Chemistry
Sofia, Bulgaria
Chapter 6

Shrikant B. Dhoot
Department of Chemical Engineering
University Institute of Chemical Technology
University of Mumbai
Matunga, Mumbai, India
Chapter 8

Adriana Gushterova
Institute of Microbiology
Bulgarian Academy of Sciences
Sofia, Bulgaria
Chapter 6

Ülkü Mehmetoğlu
Department of Chemical Engineering
Ankara University Engineering Faculty
Ankara, Turkey
Emails: mehmet@eng.ankara.edu.tr, ulku.mehmetoglu@eng.ankara.edu.tr
Chapter 7

Peter Nedkov
Bulgarian Academy of Sciences
Institute of Organic Chemistry
Sofia, Bulgaria
Chapter 6

Maria Papagianni
Department of Hygiene and Technology of Food of Animal Origin
School of Veterinary Medicine
Aristotle University of Thessaloniki
Thessaloniki, Greece
Email: mp2000@vet.auth.gr
Chapter 3

Jonnalagadda Raghava Rao
Chemical Laboratory
Central Leather Research Institute
Adyar, Chennai, India
Email: clrichem@lycos.com
Chapter 5

Thirumalachari Ramasami
Chemical Laboratory
Central Leather Research Institute
Adyar, Chennai, India
Chapter 5

Ashwini D. Sajgure
Department of Chemical Engineering
University Institute of Chemical Technology
University of Mumbai
Matunga, Mumbai, India
Chapter 8

Ramkrishna Sen
Department of Biotechnology
Indian Institute of Technology, Kharagpur
Kharagpur, India
Email: rksen@yahoo.com
Chapter 2

Palanisamy Thanikaivelan
Centre for Leather Apparels and Accessories Development
Central Leather Research Institute
Adyar, Chennai, India
Chapter 5

Balachandran Unni Nair
Chemical Laboratory
Central Leather Research Institute
Adyar, Chennai, India
Chapter 5

Evgenia Vasileva-Tonkova
Institute of Microbiology
Bulgarian Academy of Sciences
Sofia, Bulgaria
Email: evaston@yahoo.com
Chapter 6

Antonio Villaverde
Department of Microbiology
Institute for Biotechnology and Biomedicine
Autonomous University of Barcelona
Bellaterra, Barcelona, Spain
Email: avillaverde@servet.uab.es
Foreword

Ganapati D. Yadav
Department of Chemical Engineering
University Institute of Chemical Technology
University of Mumbai
Matunga, Mumbai, India
Emails: gdyadav@udct.org, gdyadav@yahoo.com
Chapter 8

FOREWORD

Although we are used to observing them separately, enzymes act in living beings as extremely complex, network-like mixtures that support all the biochemical transformations on which life is based. In the biotechnological context, many of the enzymatic processes performed in vitro at both small and industrial scales lie on the enzymatic transformation of a single molecular species for the generation of a product and are catalyzed by a single enzyme. However, the number of technological applications for which cell-free enzyme mixtures are required is increasing, and the science of how to combine individual reactions in complex processes is under speedy development. Obviously, any of the current in-progress multi-enzyme processes attempts to mimick the complexity of a living cell or cell community. However, the refined combination of selected enzymes and substrates offers a new technological approach that is supporting the development of new or improved products in many fields such as food, leather and pharmaceutical industries.

In this book, an ensemble of examples is provided to illustrate the diversity of approaches and applications to which the multi-enzyme catalysis is currently applied.

The chapter by Sen provides a general overview about fundamentals in enzymology and catalysis followed by a revision of research and review articles relevant to subspecialty applications. This introductory article describes the utilization of enzymes derived from microbes and from other type of cells for industrial exploitation.

Papagianni and Belafi-Bako have provided overviews on enzymes and enzyme mixtures in food technology. The chapter by Dr. Papagianni describes the new technologies that employ enzymes in food processing.

Chapters by Rao and Vasileva-Tonkova highlight the utility of enzymes in leather processing. Worldwide utilization of finished leather has increased significantly. Both Rao's and Vasileva-Tonkova's group are well-known for their work on enzymes in the leather industry, and these two articles provide insight into the utilization of enzymes in different leather processing steps. At Central Leather Research Institute (CLRI), Rao's group has utilized substrate specific enzyme mixtures to replace toxic and dangerous chemicals to provide an integrated solution to the pollution problems in leather processing.

Mehmetoğlu and Yadav present the utilities of enzymes in the manufacturing of fine chemicals. The chapter by Mehmetoğlu describes the production of enantiomeric compounds, and this gives an outlook on how enzymes can replace traditional organic synthesis. Such enzymatic processes are environmentally friendlier than organic synthesis and often more economical as well. Dr. Yadav has reviewed many of these processes. For a beginner it provides a good account, but the chapter also provides intriguing elements targeted to specialized researchers derived from the extensive experience of Dr. Yadav both in academia and industry.

The last chapter by Algeciras and Bhattacharya reviews the nascent attempt to recycle carbon dioxide in a concatenated and continuous fashion utilizing enzymes. Bhattacharya is at the forefront of proponents of pollution abatement at source. He has been working on the contained handling of carbon dioxide pollution resulting in useful chemicals utilizing enzymes, and this article provides a comprehensive account of collective effort in this area.

Overall this book provides a comprehensive review of the utility of enzymes and enzyme mixtures in different spheres of human activity. Both beginners and specialists will find its reading useful, inspiring and encouraging.

Antonio Villaverde
Professor of Microbiology
Institute for Biotechnology and Biomedicine
Autonomous University of Barcelona
Bellaterra, Barcelona, Spain

Acknowledgements

This book embodies the work of many authors who have contributed chapters. They and coworkers in their laboratories deserve the credit for this book. I would like to thank Professor Manuel J.T. Carrondo, Instituto de Biologia Experimental e Technologica, Portugal and John G. Aunins, Merck and Company, USA who made me look beyond the narrow scope of utilization of enzymes to broaden my horizon. I would like to thank Professor O.P. Malhotra and G.C. Baral, Banaras Hindu University, Dr. D.S. Pradhan, Biochemical division, Bhabha Atomic Research Center, Bombay, Drs. P.K. Ranjekar and Vidya Gupta, National Chemical Laboratory, Pune, Prof. P. Ghosh, Indian Institute of Technology, India who introduced me to the world of enzymes, chemical reaction engineering and biochemical engineering. I thank Drs. Vivien Yee and Foco van dan Akker for training me in structural biology of proteins and X-ray crystallography in particular and for their occasional suggestions. My friends and colleagues Professor Antonio Villaverde, Dr. James Gomes and Dr. Richard Lee and whoprofoundly helped me to shape the ideas into the form of a book. Lastly, I thank Mabel Algeciras for providing help to put all the pieces together. I am deeply grateful to my friends and colleagues for their faith in this and every other project that I have undertaken. Cynthia Conomos and Ron Landes, who generously shared their wisdom and insight, were very helpful as well. Finally, my wife Sumana Bhattacharya has been a pillar of strength during the long hours I worked on this book. I will always be grateful for her help and support.

CHAPTER 1

Introduction to the Use of Complex Enzyme Mixtures

Sanjoy K. Bhattacharya*

Abstract

Biocatalysts produce substances that are often advantageous compared to traditional methods. They also help synthesize new compounds. To achieve increased or novel production, mutant organisms, tailored enzymes as well as novel combinations of enzymes in reactors are being used and the sphere application of enzymes is expanding. The complex biosynthesis, once mainly utilized only by biopharmaceuticals, has now been embraced by food, textile and leather industries and in environmental pollution control. The new and environmentally friendly ways of manufacture are made possible by enzymatic conversion processes. Increased understanding of the functional interaction of proteins and protein-protein network is expected to alter in vitro biosynthesis in reactors.

Introduction

Biotechnological applications were once mainly used for production of biopharmaceuticals. However, it has expanded to food, chemical, leather and environmental applications.[1-3] The biosynthesis using microorganisms requires a balance between toxicity on metabolism and product yield. The yield of product is influenced by toxicity and metabolism burden that production poses on the host organism. Utilization of enzymes rather than host organism is often a compromise between purification of enzymes versus obtaining product with relatively high purity. Often a better yield of product is achieved utilizing enzymes than by an organism.[4]

Reduction of Process Steps and Increased Product Yield

Traditionally random mutagenesis has been used for improvement of production or for production of new compounds. Methods (UV radiation, chemical mutagenesis) do not require knowledge of the genome of the organism. They are, however, very labor-intensive and require rigorous screening for isolation of suitable mutants. Site-directed mutagenesis requires knowledge about the genome and metabolic pathway in the organism. Increasing productivity and reducing conversion steps improving purity or enrichment of enantiomers, improving solubility of reactants and eliminating utilization of hazardous substances are desired goals for adopting complex enzymatic biosynthesis. Elimination of toxic intermediates and high temperature and pressure is also desired. In the food and pharmaceutical industries great advantages in process economics have been achieved by such changes. Adopting enzymes for synthesis of oligosaccharides and antibiotics regio- and stereo selectivity have been achieved while reducing the number of steps.[5] Combinatorial use of enzymes has shown to result in synthesis of dTDP-4-keto-6-deoxy-D-glucose

*Sanjoy K. Bhattacharya—Bascom Palmer Eye Institute, McKnight Bldg., 1638 NW 10th Avenue, Suite 706A, University of Miami, Miami, Florida 33136, U.S.A.
Email: sbhattacharya@med.miami.edu

Enzyme Mixtures and Complex Biosynthesis, edited by Sanjoy K. Bhattacharya.

resulting in elimination of difficult synthetic steps and reduced synthesis cost.[6] Several chiral compounds are now industrially manufactured using enzymes.[7] A number of enzymes are used in organic solvents facilitating their use for sparingly soluble substances, Penicillin-G and 7-ADCA and examples of such syntheses.[8,9]

Synthesis of Novel Products and New Applications

For production of novel compounds, several approaches have been attempted including feeding an uncommon precursor, utilization of co-culture of microorganisms and genetically altered microorganisms. An example of novel compound production is polyketide biosynthesis where all of these strategies have been attempted.[10-12] A novel compound, 1-dethia-3-aza-1-carba-2-oxacephem, was synthesized using a hybrid combination involving enzymatic and chemical synthesis.[13] A wider and diverse market is expected to reduce revenue fluctuations in the biotech industry.[14] In comparison to applications is from physical sciences, several industrial biotechnological processes are still under development. Human activities and unprecedented population have necessitated expansion of habitable land and posed geological problems. An area of application yet to be explored is the use of biologically synthesized materials acting as geotextiles for land stabilization.[15] Innovative interdisciplinary approaches are the cornerstone of physical sciences which is yet to be fully embraced by the biotech industry. Application planners are yet to exploit the full potential of biocatalysis. Chemical and leather industries have adopted enzymatic approaches for production of fine chemicals and for leather processing. The former includes environmental friendly synthesis of the chiral and enantimeric compounds utilizing enzymes. Global carbon dioxide pollution is one such problem where new processes of enzymatic conversions holds promise. An enzymatic process of continuous but contained carbon dioxide fixation with acceptor recycling is possible with stationary emissions sources.[16]

Exploitation of New Information for Product Generation

Advancement in genomics and proteomics resulted in treating biological organisms as complex systems.[17] All complex systems (biological or man-made) have similarity in networks, many underlying patterns of metabolic networks share common themes[18] and obey a few simple rules that they are scale-free, clustered, possess modularity and have hierarchical organization.[18-20] The biological systems are highly compartmentalized with internal channeling that supports extraordinarily high rates of chemical conversion.[21] The channeling also supports conversion of sparingly soluble substances at physiological rates.[19,22-24]

Cells have been used for chemical conversion of small molecules where the inherent pathway within the organisms enables such conversions. The metabolic needs of the organism are often viewed as wasteful and purification of desired chemicals is cumbersome and expensive. It is unclear, how, amidst myriads of reactions, enzyme processes simultaneously maintain their specificity and how different proteins recognize their interacting partners so specifically within the cytoplasm. A group of interacting proteins within a cell are named as interactome. The protein complexes are now regarded as extreme forms of relatively highly stable interactomes[24] and are often responsible for channeling of intermediate compounds resulting in high conversion rates and the quick turnarounds triggered by an external stimulus in biological systems. Enzymes have traditionally been used in vitro in linear fashion but use of combinatorial parts of pathways or knowledge-based integration of interactomes will be advantageous to increase productivity or to produce novel compounds.[24] One advantage of such integration is to stem reverse catalysis in response to concentration by the enzymes. Expanded understanding of metabolic networks and protein-protein interaction will help in designing better reactors. The advent of genomics and proteomics has met with commensurate advancement in computational abilities. A number of tools have become available for processing the information to aid with new design and development; several databases, including BRENDA database, which contains comprehensive information on enzymes, are now available.[25] Such databases and computational tools empower in silico trials to find possible pitfalls before embarking on actual "wet" experiments.

Conclusions and Future Outlook

Enzymatic biosyntheses are nontoxic and nonhazardous as opposed to chemical synthesis methods. Novel products have been generated as a result of the different combination of enzymes often from different systems or organisms or by using different starting material in a given biosynthetic pathway. In the future, all combinations, including random and directed mutagenesis and utilization of novel starting material, will continue to be used for novel chemical production. The future expansion of complex biosynthesis will involve expanding into environment and geology in addition to food, fine chemicals, leather and pharmaceuticals. Embracing new interdisciplinary approaches will help generate unexplored utility. Such endeavors will also widen the scope for investment and safer manufacturing for several fields such as the leather and chemical industries where conventional routes are still predominant. Exploitation of metabolic networks and protein-protein interaction in new ways will bring radical changes in design as well as result in new applications.

Acknowledgement

I thank Professor Antonio Villaverde, Universitat Autònoma de Barcelona, for critical reading of the manuscript.

References

1. Kirk O, Borchert TV, Fuglsang CC. Industrial enzyme applications. Curr Opin Biotechnol 2002; 13(4):345-351.
2. Oppermann-Sanio FB, Steinbuchel A. Occurrence, functions and biosynthesis of polyamides in microorganisms and biotechnological production. Naturwissenschaften 2002; 89(1):11-22.
3. Rieger PG, Meier HM, Gerle M et al. Xenobiotics in the environment: Present and future strategies to obviate the problem of biological persistence. J Biotechnol 2002; 94(1):101-123.
4. Velkov T, Lawen A. Nonribosomal peptide synthetases as technological platforms for the synthesis of highly modified peptide bioeffectors—Cyclosporin synthetase as a complex example. Biotechnol Annu Rev 2003; 9:151-197.
5. Perugino G, Trincone A, Rossi M, Moracci M. Oligosaccharide synthesis by glycosynthases. Trends Biotechnol 2004; 22(1):31-37.
6. Oh J, Lee SG, Kim BG et al. One-pot enzymatic production of dTDP-4-keto-6-deoxy-D-glucose from dTMP and glucose-1-phosphate. Biotechnol Bioeng 2003; 84(4):452-458.
7. Patel RN. Microbial/enzymatic synthesis of chiral pharmaceutical intermediates. Curr Opin Drug Discov Devel 2003; 6(6):902-920.
8. Ulijn RV, De Martin L, Halling PJ et al. Enzymatic synthesis of beta-lactam antibiotics via direct condensation. J Biotechnol 2002; 99(3):215-222.
9. Schroen CG, Nierstrasz VA, Bosma R et al. Process design for enzymatic adipyl-7-ADCA hydrolysis. Biotechnol Prog 2002; 18(4):745-751.
10. Weber T, Welzel K, Pelzer S et al. Exploiting the genetic potential of polyketide producing streptomycetes. J Biotechnol 2003; 106(2-3):221-232.
11. Watanabe K, Khosla C, Stroud RM, Tsai SC. Crystal structure of an Acyl-ACP dehydrogenase from the FK520 polyketide biosynthetic pathway: Insights into extender unit biosynthesis. J Mol Biol 2003; 334(3):435-444.
12. Mazur MT, Walsh CT, Kelleher NL. Site-specific observation of acyl intermediate processing in thiotemplate biosynthesis by fourier transform mass spectrometry: The polyketide module of yersiniabactin synthetase. Biochemistry 2003; 42(46):13393-13400.
13. Hakimelahi GH, Li PC, Moosavi-Movahedi AA et al. Application of the Barton photochemical reaction in the synthesis of 1-dethia-3-aza-1-carba-2-oxacephem: A novel agent against resistant pathogenic microorganisms. Org Biomol Chem 2003; 1(14):2461-2467.
14. Strock WJ. Stocks rise in fourth quarter. Indexes for chemical drug and biotech firms were up from third quarter and from 2002. Chem Eng News 2004; 82:21-22.
15. Sukias JP, Craggs RJ, Tanner CC et al. Combined photosynthesis and mechanical aeration for nitrification in dairy waste stabilisation ponds. Water Sci Technol 2003; 48(2):137-144.
16. Bhattacharya S, Chakrabarti S, Bhattacharya SK. Bioprocess for recyclable CO2 fixation: A general description. In: Bhattacharya SK, Chakrabarti S, Mal TK, eds. Recent Research Adv in Biotechnol Bioengg. Trivandrum, Kerala: Research Signpost; 2002:109-120.
17. Bialek W, Botstein D. Introductory science and mathematics education for 21st-Century biologists. Science 2004; 303(5659):788-790.

18. Bhattacharya SK. Enzyme facilitated solubilization of carbon dioxide from emission streams in novel attachable reactors/devices. 2003; US patent application serial 464789 No. 20050214936.
19. Spirin V, Mirny LA. Protein complexes and functional modules in molecular networks. Proc Natl Acad Sci USA 2003; 100(21):12123-12128.
20. Cho S, Park SG, Lee do H, Park BC. Protein-protein interaction networks: From interactions to networks. J Biochem Mol Biol 2004; 37(1):45-52.
21. Nagradova N. Interdomain communications in bifunctional enzymes: How are different activities coordinated? IUBMB Life 2003; 55(8):459-466.
22. Schneider K, Hovel K, Witzel K et al. The substrate specificity-determining amino acid code of 4-coumarate:CoA ligase. Proc Natl Acad Sci USA 2003; 100(14):8601-8606.
23. Morgan JA, Clark DS. Salt-activation of nonhydrolase enzymes for use in organic solvents. Biotechnol Bioeng 2004; 85(4):456-459.
24. Chakrabarti S, Bhattacharya S, Bhattacharya SK. Biochemical engineering: Cues from cells. Trends Biotechnol 2003; 21(5):204-209.
25. Schomburg I, Chang A, Ebeling C et al. BRENDA, the enzyme database: Updates and major new developments. Nucleic Acids Res 2004; 32(Database issue):D431-433.

Chapter 2

Metabolic Engineering in the Targeted Improvement of Cellular Properties in Plants vis-à-vis Biopharmaceutical Production

Ramkrishna Sen*

Abstract

Modern biotechnology has evolved from genetic engineering to introduce single gene traits to metabolic engineering to manipulate multigenic traits, thereby coding for complete metabolic pathways, bacterial operons, or therapeutic molecules that require an assembly of complex multi-subunit proteins. Mastering metabolic engineering for directed improvement of biological cell factories by manipulating enzymatic, transport and regulatory functions with the use of r-DNA technology is the essence and hallmark of the production of recombinant proteins for human therapy, popularly known as biopharmaceuticals. Metabolic engineering in plants is currently at the threshold of an exciting new paradigm which emphasizes the introduction of traits that need manipulation of metabolic pathways or coordinated expression of multi-subunit proteins. Orchestrating the various ways of controlling gene expression in transgenic plants and of the numerous techniques of developing a better understanding of the fundamental physiology of the process, the expression level, timing, subcellular location, and tissue or organ specificity is what makes metabolic engineering in plants so challenging and exciting. It promises to create new opportunities in custom designing and delivering novel molecules for therapeutic and biomedical applications. Thus, this chapter focuses on the tremendous potential of plants as cost-effective and sustainable multi-gene expression platform and transgenic manufacturing systems and discusses both the current status and role of functional genomics and metabolic engineering in the targeted improvement of cellular activities for the large-scale production of biopharmaceuticals.

Introduction

The new millennium has ushered in an era of biotechnology missions driven by the huge pool of information and knowledge generated by the human genome project molecular diagnostics using recombinant gene product as well as gene-targeted therapy are leading towards personalized medicines. Research endeavors worldwide are focusing on delineating the multi-enzyme biosynthetic pathways and targeting the enzymatic steps, mainly the rate-limiting ones, at the molecular level by using metabolic flux balance and control analyses for redirecting the metabolic fluxes towards enhancing the yields of the desired gene products and the bioprocess performance.[1-3] While both metabolic engineering tools are being used to determine the targets, engineering, silencing or knocking out the targeted genes and PCR-based site-directed mutagenesis applied molecular biology techniques are gaining importance in manipulating multi-genes to engineer an entire metabolic pathway or to develop overproducing mutants.[1,2] Therefore, it is of paramount

*Ramkrishna Sen—Department of Biotechnology, Indian Institute of Technology, Kharagpur, West Bengal 721302, India. Email: rksen@hijli.iitkgp.ernet.in

Enzyme Mixtures and Complex Biosynthesis, edited by Sanjoy K. Bhattacharya.

importance to comprehensively understand metabolic pathways involving enzyme mixtures leading to complex biosynthesis of metabolites and regulatory proteins in natural and genetically engineered plant, animal and microbial cells.[1-5] This chapter addresses issues related to the production of biopharmaceuticals in plants involving the integration and regulation of complex catabolic and biosynthetic processes mediated by multi-enzyme systems for the maintenance of balanced metabolism in which energy generation and precursor production are harmonized and also discusses the role of metabolic engineering in elucidating the targeted improvement of cellular properties in transgenic plants, a cheaper and sustainable heterologous protein expression platform and manufacturing system.

Cellular Metabolism

Central to our understanding of cellular metabolism, irrespective of the type of living system (microbe/plant/animal), is the experimentally validated knowledge that pathway intermediates **are the starting materials for biosynthetic processes.** These intermediates are energy (ATP), reducing power (NADH or NADPH) and small precursor molecules (acetyl coA, pyruvate, ribose-5-phosphate etc.) resulting from fuelling reactions (oxidation).[4] The other important and mandatory components of cell metabolism include a multi-enzyme system or enzyme mixtures for biocatalysis, nitrogen and sulfur for biosynthesis of regular and sulfur-containing amino acids, nucleotides and sulphonated carbohydrates, enzyme cofactors including some metal ions and prosthetic groups, and trace elements. The diagram in Figure 1 illustrates the relationship between the products of the fuelling reactions and the building blocks for cellular macromolecule synthesis.

It is puzzling to imagine how the tiny cells efficiently integrate and economically regulate their fuelling and biosynthetic reactions to produce the twelve precursor molecules in correct proportions and, hence, the building blocks for new cell synthesis in a balanced and sustainable fashion (Fig. 2). It is equally intriguing to observe that the cells strategically coordinate the intricate milieu and network of catabolic and anabolic reactions—even when faced with adverse circumstances such as change in the availability of nutrients and energy—through the use of coupling agents and by controlling the material and energy exchange fluxes through particular pathways and the consequential flow of genetic information (Fig. 3). For all cells, ATP as a coupling agent in energy metabolism is central to metabolic regulation and integration. Other chemical coupling agents include the activated metabolic intermediates such as NADH, NADPH, $FADH_2$ and Nucleoside tri-phosphates etc that are usually assigned a value in terms of ATP equivalents. **An interesting point concerning metabolic regulation and integration is that the activation of key enzymes or enzyme mixtures govern the rates of biosynthetic pathways more than mass action effects.**[4-5] It is thus pertinent to discuss and illustrate the targeted improvement of cellular properties and the productivity of novel compounds by metabolically engineering the multi-enzyme pathways.

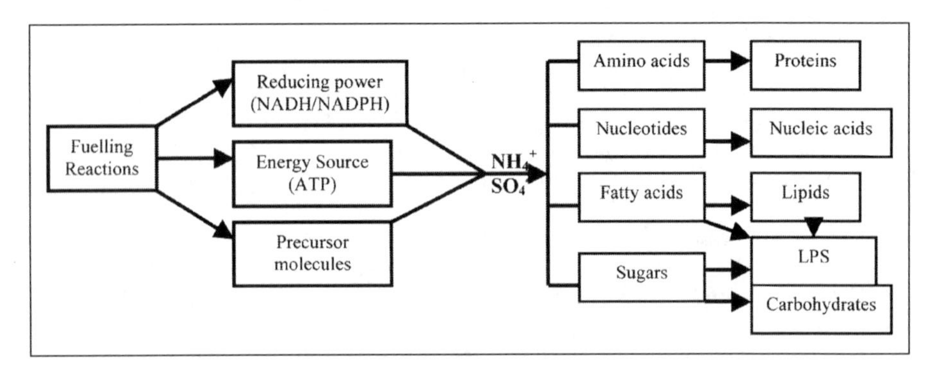

Figure 1. Components in driving biosynthesis from fuelling reactions to the building units for cell synthesis (LPS = lipopolysaccharides). Adapted from reference 4.

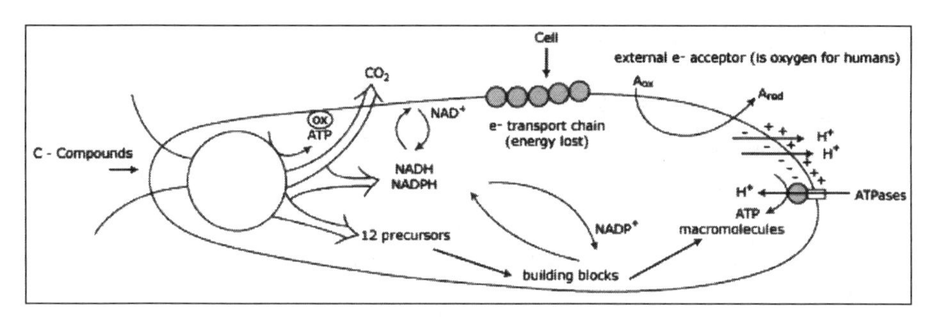

Figure 2. A schematic representation of the cell metabolism in chemoheterotrophs.

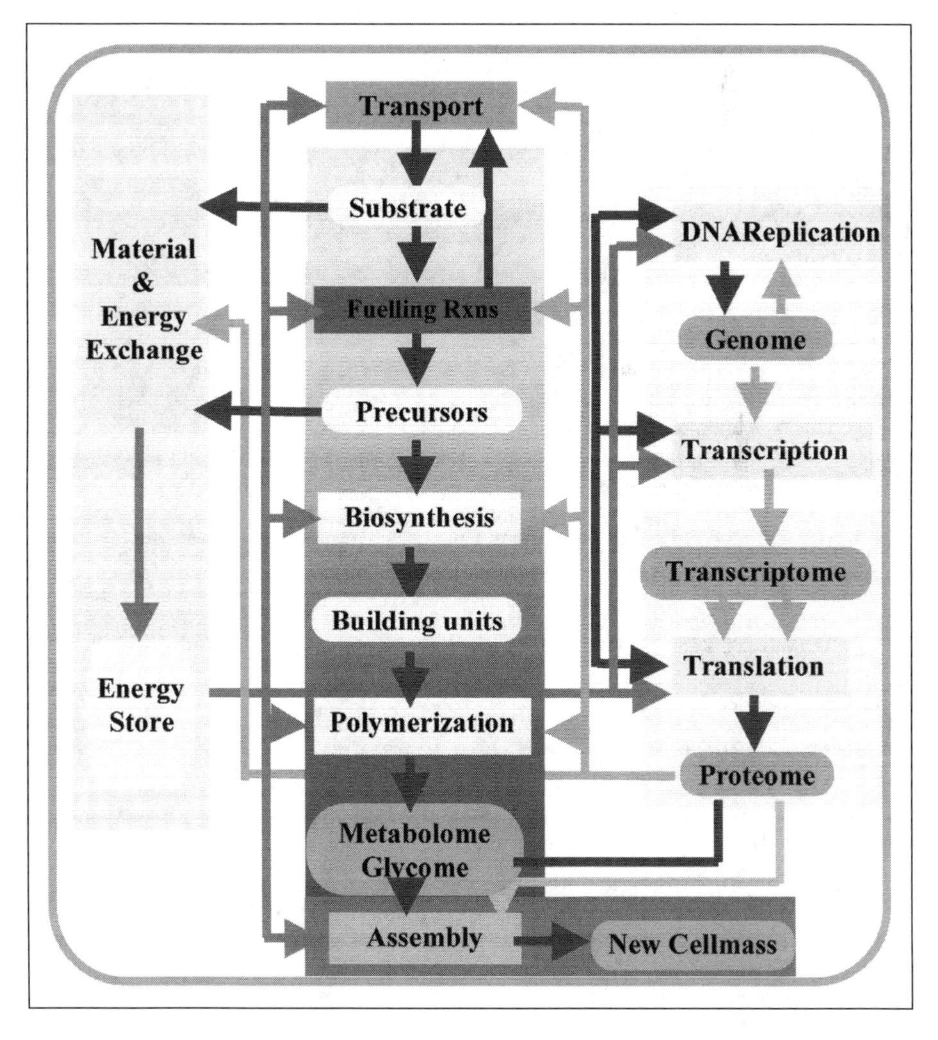

Figure 3. Cellular coordination of a biochemical reaction network, material and energy fluxes and the flow of genetic information.

Metabolic Pathway Engineering

Metabolic Engineering of Biological Systems

While trying to manipulate cellular properties at the genetic level to our advantage, traditional mutagenesis could not endow a cell with genes from other organisms.[6] Recombinant DNA technology made such work possible. For the first time, by expressing new combinations of activities in cells, by silencing or knocking out certain gene targets, or by carrying out site-directed, PCR-based mutagenesis it has been possible to induce cell factories to synthesize products of commercial interest.[6-7] The outcomes of these advanced techniques have been mathematically explained by stoichiometry and molecular modeling, material and energy balances, metabolic flux analysis and control strategies as well as by thermodynamic interpretations in silico. The outcomes have also been experimentally validated in vitro under the common umbrella of metabolic or cellular engineering[1-3] which has proven to be an effective and unique route to the biosynthesis of new complex chemicals: novel antibodies, polyketides, glycoproteins, biopolymers, and many more.

Microbes have been the natural targets for manipulation for the first wave of metabolic engineering development because of their short generation times, easy handling in bioreactors and, of course, well elucidated genetic information and standardized tools, thus accelerating experimentation, process and product development.[6] These properties also make microbes, mainly bacteria, a preferred model system for mathematical and experimental validation and design. Thus, microbial biosynthetic pathways have been the most attractive targets for metabolic engineering because of the potential for high-leverage effects on production rates and simpler regulation than more robust central metabolism thereby reducing the probability of unexpected cellular responses.[1-2,6-7] New secondary metabolite synthesis in microbes that introduce target genes and hence, pathways that generate novel chemical entities have constituted the challenges of metabolic engineering in the last decade.[7] Metabolic engineering of plant and animal cells is of immediate interest for biotherapeutic protein production and modification of the basic cell platform, in terms of growth and continuous product synthesis characteristics, has emerged as a new challenge and fresh source of opportunities, as has post-translational modification by glycosylation to provide novel, more therapeutically effective glycolproteins.[6] Table-1 summarizes some commercially important types of biotechnology products produced by transgenic biological systems, some of which are the results of well-executed metabolic engineering strategies.

A relative comparison between major heterologous protein expression platforms and the utility of transgenic or recombinant manufacturing systems has been shown schematically in Figure 4. Though glycosylation and other eukaryotic post-translational modifications such as proper folding or assembling and localization specifically demand the use of animal or plant cell platforms for the large-scale production of recombinant biotherapeutics, engineering the metabolic pathways appears to be more challenging in these advanced and intricate systems than in relatively simple prokaryotic expression platforms.

Table 1. Transgenic bio-systems as factories for commercially viable products

Type of Products	Examples
1. Bio-pharmaceuticals	Vaccines, antimicrobials, therapeutic enzymes, antibodies, cytokines, growth factors etc.
2. Industrial chemicals and enzymes	Solvents (alcohols, acetone etc.), modified starch, cellulase, lipase, amylase, protease etc.
3. Animal feed/food	Probiotics, phytase, xylanase, single cell proteins and oils, etc.
4. Advanced/specialty products	Biopolymers (Polyhydroxy alkanoates, Polylactic acid, silk, gelatin), biosurfactants, bioinsecticides, biofertilizers, etc.

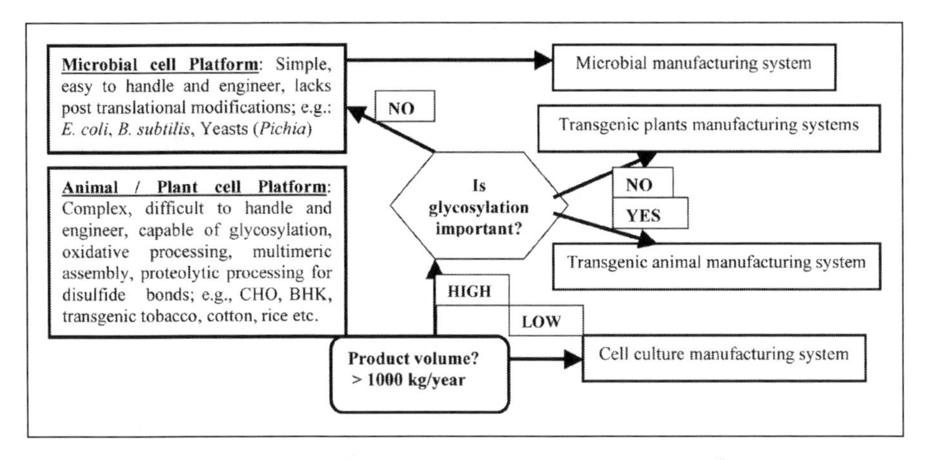

Figure 4. Recombinant protein expression platforms and transgenic manufacturing systems.[9]

Biopharmaceuticals: A Perspective

Modern biotechnology has now evolved from an initial phase of genetic engineering to introduce single gene traits to the phase of metabolic engineering for manipulating multigenic traits, thereby coding complete metabolic pathways, bacterial operons or therapeutic molecules that require an assembly of complex multi-subunit proteins. Mastering the technology that involves such metabolic engineering to direct improvement of the biological cell factories performance by manipulating enzymatic, transport and regulatory functions with the use of r-DNA technology, is the essence and hallmark of biosynthesizing and isolating recombinant proteins for human therapy, popularly known as biotherapeutics/biopharmaceuticals.[7-8] The most popular blockbuster biopharmaceuticals include filgrastim (GCSF) and erythropoietin which represent the fields of oncology and hematology. Developing or acquiring the technology to obtain these and other FDA-approved biopharmaceuticals such as α and γ interferons, interleukins, alteplase, pegaspargase etc, forms the basis to enter the most demanding healthcare markets, since it surmounts the entry hurdles in the field of technology, patent protection and registration. Table 2 compares the efficacies and production profiles of the traditional drugs and biopharmaceuticals.

Table 2. Biopharmaceuticals versus traditional drugs: A comparative statement

Traditional Drugs	Biopharmaceuticals
1. Nonspecific binding to the target receptors	1. Highly specific binding to the receptors
2. Possible interactions with other drugs, causing deleterious side effects.	2. Rare possibility of drug interactions
3. Possible carcinogenic activities	3. Not carcinogenic
4. Complex pharmacokinetics, lower degree of bio-equivalence and bioavailability	4. Easy pharmacokinetics as breakdown is predictable, relatively higher degree of bio-equivalence and bioavailability
5. Theoretically, any target molecule can be reached.	5. Target molecules are restricted, to only the outside of the cells
6. Rare occurrences of immune reactions	6. Possible immunogenic effects
7. Success rate 6% in phases I-III of clinical trials.	7. Success rate 20-30% in phases I-III of clinical trials
8. High drug development costs, but low production costs	8. Relatively lower development costs and higher production costs.

Table 3. Selected biopharmaceuticals and their global sales figures (2003)

Biopharamceutical	Manufacturer	Disease Symptoms	Sales Value (US$, millions)
Humulin (Insulin)	Eli Lilly	Diabetes	1060
Humalog (Insulin)	Eli Lilly	Diabetes	1020
Epogen and Aranesp (both EPO products)	Amgen	Anaemia (cancer and kidney failure)	3978
Neulasts and Neupogen (both EPO products)	Amgen	Anaemia (cancer and kidney failure)	2522
Procrit / Eprix (EPO)	Johnson & Johnson	Anaemia	3984
Avonex (IFN-β)	Biogen	Multiple sclerosis	1188
Remicade (MAb)	Johnson & Johnson	Cohn's disease	1729
Rituxan (MAb)	Genentech	non-Hodgkin's lymphoma	1489
Herceptin (MAb)	Genentech	Breast cancer	425
Humira (MAb)	Abbott	Rheumatoid arthritis	250
Enbrel (Etanercept)	Amgen	Rheumatoid arthritis	1300
Filgrastim and Prefilgrastim (both GCSF products)	Amgen	Myeloid leukaemia	664 (in first qtr.)

EPO: Erythropoietin; MAb: Monoclonal antibody; IFN: Interferon; GCSF: Granulocyte colony stimulating factor. Adapted from Walsh, 2005 (ref. 8) and company websites.

With the mapping of the human genome and the concomitant explosion of proteomics, a steady stream of biopharmaceuticals has been launched: recombinant therapeutic proteins, monoclonal antibody-based therapeutic/in vivo diagnostic products and nucleic acid-based products. Biopharmaceuticals now represent approximately one in every four genuinely new pharmaceuticals (new molecular entities, NMEs) coming on the market.[8] Examples of such blockbuster drugs and their sales values are shown in Table 3.

With the advent of functional genomics, the timeframe for drug discovery from the identification of gene to the identification and synthesis of lead molecules has drastically reduced from three to seven years to one to two years (Fig. 5).

Advances in functional genomics, high-throughput screening and evolution technologies have also resulted in the development and increased availability of new and robust biocatalysts suited for industrial-scale applications, particularly in the synthesis of enantiomerically pure molecules as pharmaceutical intermediates.[10] Tailor-made enzymes or enzyme mixtures, produced by mutants and engineered organisms utilizing knowledge of systems biology or by combinatorial means, can be used to increase existing biopharmaceutical production or to synthesize novel products in reactors.[5] Though the journey from genetic engineering to protein engineering has been smooth to produce some engineered antibodies, and in spite of intensive R&D initiatives and efforts, no product produced in a transgenic-based system has yet gained regulatory approval. Recently, GTC Biotherapeutics (http://www.transgenics.com) applied to Europe's regulatory authority (EMEA) for obtaining marketing authorization for 'ATryn', a recombinant human antithrombin produced in the milk of transgenic goats. 'CaroRX', a transgenic plant-based IgG product by Planet Biotechnology (http://www.planetbiotechnology.com) is in Phase II clinical trials.[8-9] This plant-produced antibody is specific for Streptococcus mutants, the causative agent of bacterial tooth decay.

Plants as Factories for Biopharmaceuticals

Scientists dream of a world where any protein, either naturally occurring or man made, can be produced safely, inexpensively and in required quantities, by using only natural resources, i.e.,

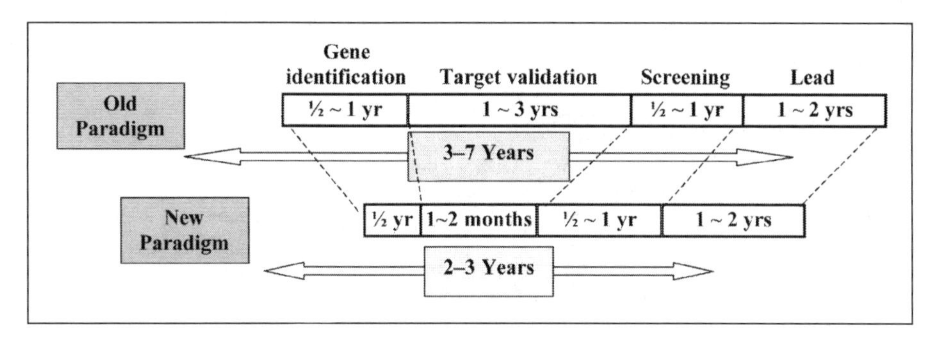

Figure 5. Functional genomics revolutionizing drug discovery (not up the scale).

available nutrients, water and sunlight. This world may soon come true as we learn to harness the power of plants to produce recombinant proteins on an agricultural scale.[11] Molecular farming in plants has already successfully produced recombinant therapeutic proteins that are approaching commercial approval and many more are expected to follow. Recent advances in the so-called arena of 'omics', namely, genomics, proteomics and metabolomics have led to the identification of new genes and novel biopharmaceuticals. As more such proteins are characterized, more biopharmaceutical targets will be identified for drug discovery. Some 160 protein-based biopharmaceuticals have gained medical approval with several hundred more in the pipeline.[8] As more protein products are developed, apprehension will be growing about the ability of the biopharmaceutical industry to meet production demands. This apprehension has initiated a debate on the selection and use of an efficient protein expression platform for the continuous and cost-effective production of such protein drugs. This concern has also brought about a change in the attitude of the old-order biotech industry towards plants as protein expression platforms and transgenic manufacturing systems. The motivations for choosing transgenic plant manufacturing system stem from the newly emerging market demands for antibodies for preventive, therapeutic and diagnostic applications; blood proteins and factors for diverse uses; oral and edible vaccines; nationwide stockpiling of biodefensive agents; remedies for chronic conditions needing multiple expensive drugs such as for the treatment of arthritis, AIDS, cancer, Alzheimer's disease etc; biogenerics and oral protein and probiotic-based nutraceuticals and functionalized protein-based biomaterials.[9,12-17] Plants as factories for therapeutic proteins have come of age and of the 99 most frequently prescribed U.S. drugs, about a fifth come from plants, from birth control pills (Mexican yam) to the anticancer drug taxol (Pacific Yew tree).[9,12-16] The major reasons for their use and some of the attractive features of plant-derived proteins are listed in Table 4.

R&D activities using of transgenic plants continue to offer the promise of commercial-scale production of safe, pure and highly efficacious biopharmaceuticals, including monoclonal antibodies (MAbs), enzymes, blood proteins and new types of subunit vaccines for preventing infectious diseases.[13,16] Biopharmaceuticals, in particular MAbs, are one of the fastest growing segments of biotherapeutics and diagnostics, with 200 products in clinical trials and many more in preclinical development. These products are useful for treating life-threatening disease conditions including arthritis, cancer, infection, inflammation and cardiovascular disease.[17] A recent survey indicated that seven plant-derived antibodies or antibody derivatives have reached advanced stages of product development[18] and about 70 therapeutic MAbs will make to the market by 2008, thereby requiring an annual manufacturing capacity of 10 metric tons of MAbs (Dry L.; www.bio.org/pmp/index.asp). If these estimates are to be believed, the result could be an unprecedented demand for huge biomanufacturing capacity for MAbs.[16] Thus, the bottom line is to focus on transgenic plants as a promising system for heterologous protein expression and post-translational modification, particularly by glycosylation of plant-made MAbs for human therapy and process scale-up.[19] Table 5 represents a comprehensive picture of the use of plants as cell factories for the production

Table 4. Advantages of plant systems for biopharmaceutical production

Important Criteria	Attractive Features
System diversity	Plant systems are more diverse due to well-understood genetics, breeding, inherent toxins and exogenous contaminants.
Protein product	High levels of accumulation possible. In most of the cases, stable product in seed can be preserved for years.
Post-translational modifications	Eukaryotic post-translational modification system, proper folding and assembling of recombinant animal proteins expressed in plants.
Protein purification	Similar purification strategies as cell culture.
Production scale	Large scale production in multiples of tons, commercially feasible.
Scale up	Simple and flexible scale up by field expansion; Hairy root culture in bioreactors for secondary metabolite production.
Cost	Low capital and production cost, in spite of longer gestation period.
Technology	Provides enabling technology for proteins, not possible from other systems. Scalable and optimized technologies possible.
Drug delivery	Easy and cost-effective delivery of lyophilized plant tissues/fruits with the protein drugs, not requiring cold-chain storage systems.

Adapted from Baez, 2004 (ref. 9).

of recombinant therapeutics and diagnostics, biopolymers, steroids etc., and also as application oriented feasible alternatives to microbial and animal cell cultures.[8,9,13]

Hence, the major products in development for human medicine using plant system-based biopharming/molecular farming are proteins—including antibodies (called 'plantibodies')—and vaccines.[19] Products like polyhydoxyalkanoates (PHA), polyhydroxybutyrate and poly-hydroxyvalerate and their copolymers can find commercial application as bioplastics and biomedical application in making scaffolds for skin grafting and wound healing.[20] Other minor but important products from transgenic plants include antimicrobials, mainly the polyketide antibiotics, nutraceutical ingredients and some cytokines and enzymes. In order to exploit fully the potential that plants offer for the production of therapeutic proteins, it might be necessary to inhibit plant-specific, post-translational modifications to get 'humanized' non-immunogenic N-glycans on Plant-made Pharmaceuticals (PMPs). The benefits that could accrue are lower manufacturing costs relative to mammalian cell culture and a reduced risk of transmission of mammalian pathogens.[12] An

Table 5. Plants as factories for recombinant therapeutics and other products

Bio-Product	Product Category	Application	Alternative To
Proteins	Antibodies	Therapeutic	Cell culture
	Vaccines	Therapeutic	Cell culture
	Cytokines/Hormones	Therapeutic	Microbial/Human
	Mammalian enzymes	Therapeutic	Animal/Human
	Blood factors	Therapeutic	Blood fractionation
	Antimicrobial peptides	Therapeutic	Animal
Biopolymers (Fatty acids or derivatives)	Polyhydroxy alkanoates	Biodegradable plastics	Microbial
	Oils	Nutra- & Cosmeceuticals	Animal
Other Metabolites	Carotenes	Nutraceuticals	Microbial
	Steroids	Nutraceuticals/Pharma	Chemical synthesis

Adapted from Baez, 2004 (ref. 9).

efficient biopharmaceutical production in plants involves properly selecting the host plant and gene expression system and requiring a decision as to whether a food crop or a nonfood crop is more appropriate.[13] Over the past few years, some notable technological advances in this flourishing area of applied biotechnology are becoming more evident from the commercial development of novel plant-based expression platforms and commendable success in solving some of the limitations of plant bioreactors such as low yields and inconsistent product quality, which create bottlenecks in the approval of PMPs.[14] Reinventing the connection between plants and human health resulted in launching a new generation of botanical therapeutics that include recombinant PMPs, multi-component botanical drugs, dietary supplements and functional foods. Most of these products will soon complement conventional pharmaceuticals in the treatment, prevention and diagnosis of diseases while at the same time adding value to traditional agriculture.[15] A critical assessment of the human and ecological risks in expressing pharmaceutical proteins in transgenic plants currently being grown in field environments has been reviewed.[16] Thus plant-based biopharmaceutical expression platform and manufacturing system represent a cost-effective and high-capacity alternative to the traditional cell culture-based production system for meeting growing demands for these biotherapeutic proteins.

Metabolic Engineering in Plants for Biopharmaceutical Production

Strategic collaborations and acquisitions are taking place worldwide to develop or acquire innovative expression systems based on plant cell cultures grown in bioreactors and plant cell lines as a general platform system for high-level expression of recombinant proteins. These systems provide the highest possible amplification level of the gene of interest and the simultaneous shut-off of the other biosynthetic processes in the cell by employing metabolic engineering approaches and tools. This results in the highest theoretically possible yield of the desired recombinant protein in the plant cell. Introduction of transgenic-biopharmaceutical-production technologies and transgenically produced human therapeutics in plants will revolutionize the healthcare market by significantly reducing the prices of these pharmaceuticals and by markedly increasing the profits for the companies, a major part of which can be used to offset or recover the huge R&D expenses that go into developing the technology. Some of the transgenic PMPs with the status of their clinical trials are listed in Table 6.

Metabolic engineering of plants promises to create new opportunities in custom designing and delivering novel molecules for therapeutic and biomedical applications[21] and also for the production of useful biochemicals like PHA, development of pest resistant, salt- and temperature-tolerant plants in agriculture and for environmental applications.[20,21] However, metabolically engineering plants by introducing and expressing foreign genes poses many technical challenges that are not encountered with microbial systems.[21] Unlike bacteria, plants cannot normally express genes from polycistronic messages and hence, requires a mechanism to coordinate the expression of complex traits involving multiple transgenes. Due to the presence of numerous organelles, not found in mammalian or yeast cells, plants need to tackle complex issues of compartmentalization of resources as well as variable extent of gene expression in differentiated organs and targeting of gene products. In addition, environmental effects may introduce an additional level of variability and uncertainty that is not encountered in bioreactor-based cell culture systems.[21] In this case, gene silencing presents an excellent opportunity to block the expression of endogenous genes.[22] Knocking out the expression of one or more endogenous genes by implanting, for example, transgenes expressing antisense RNA, self-complementary RNA, RNAi, or perhaps even just high levels of RNA is a desirable outcome in some metabolic engineering applications without the need for homologous recombination.[21,22] While making use of strong constitutive promoters to drive expression of foreign proteins in plants has proven to be useful for certain applications, heterologous expression has been found to impose too much of a burden on the host plant, causing a metabolic drain and/or impacting the plant's agronomic viability.[23] Thus, producing heterologous proteins in specific organs such as leaves or seeds might often be preferred to the general method of over-expression. Details of one such strategy for producing in plants a biopharmaceutical,

Table 6. Transgenic products with the status of clinical trial

Protein/Subunit Expressed	Plant System	Status
Hirudin (Anticoagulant)	Rapeseed	Commercially grown in Canada
Hepatitis B envelope	Tobacco	First expressed protein, third to reach clinical trial
Rabies virus glycoprotein	Tomato	First edible protein
E.coli. heat labile enterotoxin	Tobacco, potato	First to reach clinical trial Second to reach clinical trial
Diabetes autoantigen	Tobacco, potato	First protein for any autoimmune disease
Cholera toxin B-subunit	Tobacco, potato	First in chloroplast
Cholera/Rotavirus/*E.coli.* antigen fusion	Potato	First multivalent for several enteric diseases
Porcine gastroenteritis glycoprotein	Tobacco, corn	First for oral animal feeding

Adapted from Baez 2004 (ref. 9).

hirudin that has met recent commercial success in Canada, have been discussed in literature.[21,24,25] There are other methodologies that have been used to engineer the metabolic pathways in plants to produce useful products.[26,27] Some of the most expensive biopharmaceuticals of restricted availability, such as glucocerebrosidase, could become much cheaper and more plentiful through production in transgenic plants.[25] Correctly choreographing the permutations and combinations of various ways of controlling gene expression in transgenic plants and of numerous techniques to develop a clear understanding of the fundamental physiology of the process, the expression level, timing, subcellular location, and tissue or organ specificity makes metabolic engineering in plants so challenging and exciting.

Metabolic engineering in plants is currently at the cross-road of an exciting new paradigm which emphasizes the introduction of traits that need the manipulation of metabolic pathways or coordinated expression of multi-subunit proteins. An example of nuclear multigene engineering is the expression of three enzymes of the polyhydroxybutyrate (PHB) pathway.[28,30] A quadruple construct, comprising a selectable marker and three cassettes (each containing one of the three *phb* genes with a plastid targeting signal) flanked by a 35S promoter and *nos* (nopaline synthase) terminator, was used to introduce three genes involved in this pathway.[29,30] This approach resulted in a large accumulation of PHB fourfold higher than previous reports. An early success story in this new era was written by the newly developed rice varieties enriched in provitamin A. This chloroplast-based transgenic approach has facilitated expression of bacterial operons and biopharmaceuticals at unprecedented levels and resulted in the formation of insecticidal protein crystals or inclusion bodies of biopharmaceuticals with the accumulation of about 50% of foreign proteins in the total soluble protein in chloroplast transgenic plants. Foreign transcripts in transgenic chloroplasts accumulated a huge 17000% more than the best nuclear transgenic plants—an incredible achievement that not only belies concerns about gene silencing and position effects, but also eliminates the need for time-consuming breeding to bring multiple transgenes within a single host.[30]

The world business based on metabolic engineering is growing rapidly. The global sales forecast for metabolically engineered biopharmaceuticals will be around $59 billion in 2010.[31] By 2010, analysts predict up to a further 50% increase in total approved product numbers, with a total market value of Euro 52 billion, as opposed to a total value of Euro 34 billion in 2004.[32] However, in the long run, continued R&D missions on multi-gene based metabolic engineering involving the antisense, gene knock-out and RNAi centric technologies and the knowledge of functional genomics and proteomics are likely to underpin the development and approval of a more diverse range of biopharmaceuticals.

Conclusion

A great deal of the excitement over the potential of metabolic engineering in plants for the targeted improvement of cellular activities vis-à-vis biopharmaceutical production is driven by the enabling technologies of functional genomics and proteomics. These technologies have conferred us with the ability to decipher and to manipulate plants as a cheaper and sustainable heterologous protein expression platform and transgenic manufacturing system for biopharmaceuticals and also provided us with a greater understanding of gene regulation and control of plant metabolic pathways that can be engineered to our advantage with unprecedented efficiency and reliability. The metabolic engineering goal of identifying genes that confer a particular phenotype is conceptually and methodologically congruent with central issues in functional genomics. Developments in functional genomics have equipped us with new tools and approaches for understanding, mapping, modeling, and manipulating plant cells in a new paradigm. In this new era, multigene engineering will rely heavily on both nuclear- and chloroplast-centric metabolic engineering to utilize the quantum of knowledge acquired in the post-genomic era for biotechnological applications and to understand and engineer complex metabolic pathways by involving the antisense, gene knock-out and RNAi based technologies for the cost effective large-scale production of biopharmaceuticals. Thus, transgenic plants as cell factories for the production of recombinant therapeutics and diagnostics offer economically feasible alternatives to microbial and animal cell cultures.

References

1. Stephanopoulos GN, Aristidou AA, Nielsen J. Metabolic Engineering: Principles & Methodologies. San Diego: Academic Press, 1998.
2. Lee S-Y, Papoutsakis ET. The challenges and promise of metabolic engineering. In: Lee S-Y, Papoutsakis ET, ed. Metabolic Engineering. New York: Marcel Dekker, 1999:1-11.
3. Shimizu H. Metabolic engineering: Integrating the methodologies of molecular breeding and bioprocess systems engineering. J Biosci Bioeng 2002; 94:563-573.
4. The BIOTOL team. Biosynthesis and the Integration of Cell Metabolism. Oxford: Butterworth-Heinemann, 1992.
5. Bhattacharya SK. The use of enzyme mixtures for complex biosyntheses. Curr Opin Biotechnol 2004; 15:449-455.
6. Bailey JE. Introduction. In: Lee S-Y, Papoutsakis ET, ed. Metabolic Engineering. New York: Marcel Dekker, 1999:xi-xvi.
7. Bailey JE. Toward a science of metabolic engineering. Science 1991; 252:1668-75.
8. Walsh G. Biopharmaceuticals: Recent approvals and likely directions. Trends Biotechnol 2005; 23:553-558.
9. Baez J. State of the science: Role of transgenic technology in the biosynthesis of biopharmaceuticals and industrial proteins. Presentation in the risk assessment symposium of corn produced pharmaceuticals and industrial products. Iowa State University, April 2004; http://www.bigmap.iastate.edu/2004CONF/pdf/baez.pdf
10. Panke S, Wubbolts M. Advances in biocatalytic synthesis of pharmaceutical intermediates. Curr Opin Chem Biol 2005; 9:188-194.
11. Ma JK, Drake PM, Christou P et al. The production of recombinant pharmaceutical proteins in plants. Nat Rev Genet 2003; 4:794-805.
12. Gomord V, Chamberlain P, Jefferis R, Faye L. Biopharmaceutical production in plants: Problems, solutions & opportunities. Trends Biotechnol 2005; 23:559-65.
13. Goldstein DA, Thomas JA. Biopharmaceuticals derived from genetically modified plants. QJM 2004; 97:705-716.
14. Fischer R, Stoger E, Schillberg S et al. Plant-based production of biopharmaceuticals. Curr Opin Plant Biol 2004; 7:152-158.
15. Raskin I, Ribnicky DM, Komarnytsky S et al. Plants and human health in the twenty-first century. Trends Biotechnol 2002; 20:522-531.
16. Peterson RKD, Arntzen CJ. On risk and plant-based Biopharmaceuticals. Trends Biotechnol 2004; 22:64-66.
17. Fischer R, Twyman R, Schillberg S et al. Production of antibodies in plants and their use for global health. Vaccine 2003; 21:820-825.
18. Sala F, Rigano MM, Barbante A et al. Vaccine antigen production in transgenic plants: Strategies, gene constructs and perspectives. Vaccine 2003; 21:803-808.

19. Gomord V, Sourrouille C, Fitchette A-C et al. Production and glycosylation of plant-made pharmaceuticals: The antibodies as a challenge. Plant Biotechnol J 2004; 2:83-100.
20. Madison LL, Huisman GW. Metabolic engineering of Poly(3-hydroxy-alkanoates): From DNA to plastic. Microbiol Molecular Biol Rev 1999; 63:21-53.
21. Lessard PA, Kulaveerasingam H, York GM et al. Manipulating gene expression for the metabolic engineering of plants. Metabol Eng 2002; 4:67-79.
22. Kooter JM, Matzke MA, Meyer P. Listening to the silent genes: Transgene silencing, gene regulation and pathogen control. Trends Plant Sci 1999; 4:340-47.
23. Hanson AD, Shanks JV. Plant Metabolic Engineering—Entering the S Curve. Metabolic Engineering 2002; 4:1-2.
24. Parmenter DL, Boothe JG, van Rooijen GJ et al. Production of biologically active hirudin in plant seeds using oleosin partitioning. Plant Mol Biol 1995; 29:1167-1180.
25. Giddings G, Allison G, Brooks D, Carter A. Transgenic plants as factories for biopharmaceuticals. Nat Biotechnol 2000; 18:1151-1155.
26. Kumagai MH,Turpen TH, Weinzettl N et al. Rapid, high-level expression of biologically-active alpha-trichosanthin in transfected plants by an RNAviral vector. Proc Natl Acad Sci USA 1993; 90:427-430.
27. Lindbo JA, Fitzmaurice WP, della-Cioppa G. Virus mediated reprogramming of gene expression in plants. Curr Opin Plant Biol 2001; 4:181-185.
28. Bohmert K, Balbo K, Kopka J et al. Transgenic Arabidopsis plants can accumulate polyhydroxybutyrate to up to 4% of their fresh weight. Planta 2000; 211:841-845.
29. Mitsky TA, Slater SC, Reiser SE et al. Multigene expression vectors for the biosynthesis of products via multienzyme biological pathways. World patent application 2000; WO 00/52183.
30. Daniell H, Dhingra A. Multigene engineering: Dawn of an exciting new era in biotechnology. Curr Opin Biotechnol 2002; 13:136-141.
31. Datamonitor. Therapeutic proteins: Strategic market analysis and forecasts to 2010. Datamonitor (http://www.datamonitor.com) 2002.
32. Reichert JM, Pavlov A. Recombinant protein therapeutics—success rates, market trends and values to 2010. Nat Biotechnol 2004; 22:1513-1519.

CHAPTER 3

New Technologies in the Production of Food Enzymes

Maria Papagianni*

Abstract

Organisms evolve enzyme systems determined to function optimally in their natural environment for survival purposes. However, the optimal enzyme system for chemical or food processes may require characteristics that might differ considerably from those found in nature. Therefore, the greatest opportunity for improvement lies in engineering the properties that have not been naturally selected. Site-specific mutagenesis enabled biochemists to induce structural modifications on the basis of rational design factors and theoretically, no enzyme function or property is inaccessible to modification by this technology. Specific activity, pH stability, substrate-binding affinity, inhibition characteristics, are some properties which make potential targets for modification by site-specific mutagenesis. Moreover, the development over the past two decades of the techniques of genetic engineering offers a completely different approach to the selection of industrial enzymes with suitable properties. The ability to manipulate and recombine DNA allowed a more focused approach to mutagenesis, and created the potential for the manipulation of industrial microorganisms to synthesize valuable new enzymes or food ingredients. In parallel to protein engineering progress, biochemical engineering offered the alternative technology of enzyme immobilization, which aimed to significant improvements of process economics by reuse of the enzymes. However, in anticipation of a major role in industrial processing, an enormous amount of research has been targeted at new immobilization techniques, matrices and chemistries. The development of novel biocatalytic enzymes by directed evolution is a new and promising area. The number of reactions catalyzed by biological enzyme mimics, like abzymes (catalytic antibodies), is growing fast and low efficiency problems are gradually overcome. Other mimics (artificial enzymes) are also constructed on nonpolypeptide chain backbones and designed in order to withstand extreme conditions without loss of activity or conformational changes. The purpose of this chapter is to briefly review the state of the art and science of the most important new technologies involved in food enzyme engineering in a comprehensive manner.

Protein Engineering: Facing the Structure-Function Relationship Problem

An extremely promising area of genetic engineering is protein engineering. It is the application of science, mathematics, and economics to the process of developing useful or valuable proteins. Protein engineering is a relatively new discipline, with much research currently taking place into the understanding of protein folding and protein recognition for protein design purposes. New enzyme structures may be designed and produced in order to improve existing enzymes or create

*Maria Papagianni—Department of Hygiene & Technology of Food of Animal Origin, School of Veterinary Medicine, Aristotle University of Thessaloniki, Thessaloniki 54006, Greece. Email: mp2000@vet.auth.gr

Enzyme Mixtures and Complex Biosynthesis, edited by Sanjoy K. Bhattacharya.

new activities. The roles of individual amino acids in protein structure and function are beginning to be defined through site-directed mutagenesis studies of enzyme function which are fast becoming the method of choice in most laboratories, often complementing classical chemical modification studies. From the first publications on enzyme engineering were those of Winter et al[1] on tyrosyl-tRNA synthetase, Charles et al,[2] Sigal et al,[3] and Dalbaldie-McFarland et al[4] on β-lactamase. These works were mainly concerned with the methodology of mutagenesis and mutations were not designed on the basis of examination of the crystal structure. Out of these methods grew a huge body of research that has expanded enormously the understanding of protein structures.[5,6] Unfortunately from a practical point of view, much of the research effort in protein engineering has gone into studies concerning the structure and activity of enzymes chosen for their theoretical importance or ease of preparation rather then industrial relevance. This emphasis is expected to change in the future.

In protein engineering, the strategy in which the scientist uses detailed knowledge of the structure and function of a protein to make desired changes is known as "rational design". This has the advantage of being generally inexpensive and easy, since site-directed mutagenesis techniques are well-developed today. Developed by Professor M. Smith, who was awarded the Nobel Prize in 1993 for this contribution, the method has become one of biotechnology's most important instruments. Using site-directed mutagenesis, the information in the genetic material can be changed. A synthetic DNA fragment is used as a tool for changing one particular code word in the DNA molecule. This reprogrammed DNA molecule can direct the synthesis of a protein with an exchanged amino acid. An oligonucleotide-based method of site-directed mutagenesis is illustrated in Figure 1. Site-directed mutagenesis can also be achieved by using PCR (in vitro site-specific mutagenesis).

The preferred pathway for creating new enzymes by site-directed mutagenesis is by the stepwise substitution of only one or two amino acid residues out of the total protein structure. Bott et al[7] reported on the results of substituting a single amino acid for the glycine at the 166 position (in a

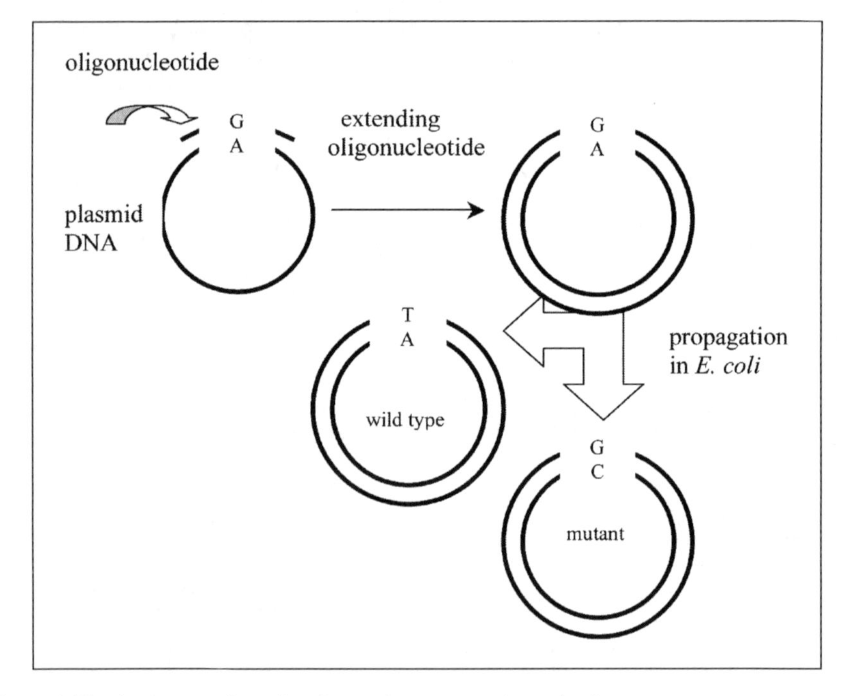

Figure 1. The basic steps in a site-directed mutagenesis method.

peptide chain of 275 amino acids) in subtilisin from *Bacillus amyloliquefaciens*. The 166 position is not part of the catalytic site, but is implicated in substrate binding. The 19 mutants examined, retained measurable enzyme activity, but differed in activity from the naturally occurring enzyme. The reported results were very indicative of the potential for the deliberate design of enzymes for specific applications. Attempts to design new substrate specificities into enzymes have met with some success as well as surprises. Certain mutational studies designed to change an enzyme specificity by substituting residues though to be involved in substrate binding, not catalysis, instead lead to a precipitous fall in enzymatic activity.[8] The reason for these, as many reported results indicate, is that science at present is at a stage where it can explain the structural consequences of amino acid substitutions after they have been determined but cannot accurately predict them.

Effective application of protein engineering necessitates a coupling of experimental data with theoretical studies for predicting the folded structure of globular polypeptides and for the development of models that anticipate the conformational effects of a given amino acid substitution. Today, a vast amount of information on protein structure is contained in numerous data banks that have emerged worldwide either commercially or governmentally sponsored. Laboratories around the world are producing immense amounts of information from DNA and protein sequences in addition to studies of kinetic, thermodynamic and structural characteristics of native and mutant proteins. Dynamic programming methods using alignment algorithms to carry out a global comparison of DNA sequences, amino acid sequences and protein structures have proved invaluable in identifying homologies and analogies provided by nature. Despite the rapid progress in data base management technology this area remains a sensitive point in protein design and attracts the efforts of engineers for the development of more easy to use and flexible algorithms.

Rational design in protein engineering becomes often limited when detailed structural knowledge of a protein is unavailable. However, the major drawback comes from the difficulty in predicting the effects of various mutations. This problem attracts today major research efforts. X-ray crystallography is a valuable tool for improving the understanding of structure-function relationship. The other valuable tool is computer modeling. Computational approaches have been applied to investigate the problem of inefficient mutagenesis, often with promising results. A recent work by Wassman et al[9] applied such an approach in predicting oligonucleotide-directed mutagenesis failures in protein engineering. Using a mehod for site-directed mutagenesis, termed "Kunkel mutagenesis", which features hybridization of a mutation-encoding oligonucleotide to a target site on a uracil-doped template plasmid, they produced results that provide a basis for improved mutagenesis efficiencies and increased diversities of cognate protein libraries.

Theoretical prediction of the structure, stability and activity of proteins would guide efficiently site-directed mutagenesis techniques. X-ray deffraction studies have provided extensive structural information for many proteins, challenging theorists to develop reliable techniques able to use such knowledge as a base for prediction of mutants' characteristics. An important work on the accurate prediction of the stability and activity effects of site-directed mutagenesis has been reported by Lee and Levitt.[10] They reported theoretical calculations of stabilization energies for 78 triple-site sequence variants of A repressor. The calculated energies correlated with the mutants' measured activities, while active and inactive mutations were discriminated with 92% reliability. They correlated even more directly with the mutants' thermostabilities, correctly identifying two of the mutants to be more stable than the wild type.

The other strategy from rational design, is known as "direct evolution." This is where random mutagenesis is applied to a protein and a selection regime is used to pick out variants that have the desired qualities. Enzymes are large and complicated catalysts. Still thousands (of ~10,000) of them, remain to be identified in terms of their three-dimensional structure. Researchers with limited resources perform random mutagenesis on a selected gene followed by expression and testing. When little is known about the structure of an enzyme, numerous mutagenized enzymes are generated rapidly and screened for changes in their characteristics. Mutants with desired characteristics are then sequenced to identify the region that determines the characteristic. The method mimics natural evolution and although a random process, it may produce superior results

Table 1. ***The most common methods applied in the process of screening and selection of mutants***

Characteristic	Selection Method
Improved yield	Use of dye-linked substrates
Improved temperature stability	Activity assays as desired temperatures
Resistance to inhibitory substances	The substance is incorporated into culture medium and enzyme activity assayed
Performance at a different pH	Using a variety of buffers and substrates enzyme activity assayed at the desired pH
Substrate accumulation	Using various carbon or nitrogen sources enzyme activity assayed for desired activity

to rational design. However, successful engineering using this strategy depends on the ability to screen for mutants. Development of suitable screens for rapid and accurate selection of mutants is central to the process of mutagenesis. Table 1 shows the most common methods applied in the process of screening and selection.

An additional technique, known as DNA shuffling, mixes and matches pieces of successful variants to produce better results. This process mimics the natural recombination that occurs during sexual reproduction. The great advantage of directed evolution techniques is that they require no prior knowledge of the structure of the enzyme, nor it is necessary to be able to predict what effect a certain mutation will have. The results of these techniques are often surprising in that desired changes are often produced by unexpected mutations. The disadvantage is that they require high-throughput, which is not always feasible. Large amounts of recombinant DNA must be mutated and the products screened for desired properties. The sheer number of variants often requires expensive robotic equipment to automate the process. Certainly, advances in high-throughput technology could greatly expand the capabilities of protein engineering.

Today, a large number of biotechnology companies develop new molecular biological methods for random and site-specific mutagenesis to create large libraries of enzyme variants. The methods applied for the enzymatic modification range from computer-aided enzymatic structure analysis and design, genetic production, biochemical characterization of the enzymes up to the structural analysis of the enzymes. Improved enzymes are then identified by suitable screening methods. Many companies provide access to libraries of various enzyme variants and conduct the screening of these libraries for the selection of biocatalysts with desired properties. Numerous alcohol-, amino acid- and hydroxyacid-dehydrogenases, and many other enzymes- libraries are examples for many companies' ranges of products.

Recombinant DNA Techniques and Food Enzymes

In the late 1970s, a set of methodologies emerged that permitted isolation of discrete DNA segments that could be inserted into living cells. The key to this method was the discovery of restriction and ligation enzymes. The ability to cut the DNA molecule, insert a new sequence, and recombine, led to a new approach, more focused compared to mutagenesis, and provided the potential for the manipulation of industrially important microorganisms to produce a whole range of products destined for the food industry.

The techniques of genetic engineering and gene cloning have been applied to food production since the 1980s. Genetically engineered bacteria and fungi are routinely used as sources of enzymes for the manufacture of a wide variety of processed foods. In a few cases, genetically engineered yeast has also been approved for food uses, but these are not yet being used in food production. Large-scale production of recombinant enzymes by bacteria or fungi takes place in high-volume bioreactors, the process followed by the downstream processing steps towards purification. The downstream processing steps separate the producing organisms and only pure enzymes take the way to the food processing industry. In most cases, the enzymes used in food manufacture cannot be traced in finished products since they are destroyed or removed by further processing and this is a reason they are rarely listed in the ingredients list.

Many of the enzymes used in food production were originally taken from non-engineered microorganisms, however suitable engineering permitted the construction of new strains of these organisms, which exhibit a number of desired characteristics such as: Increased yields from reliable sources, improved purity, and functionality suited for particular processing uses. Gene cloning is the subject of numerous published works today and the reader is redirected to them for further reading. Here a brief description will be given.

Enzymes with desired properties for certain applications are very often produced in very low levels by microorganisms unsuitable for modern fermentation processes, or in hosts unacceptable for food applications. Such enzymes make candidates for cloning into suitable production organisms (food-grade). The amino acid sequence of the enzyme-cloning candidate is determined with automatic analyzers, a process that often takes time and effort. With DNA probe construction and screening of genome libraries (pools of DNA segments that are likely to contain the gene of interest) the gene that encodes for the enzyme is identified. Transformation follows, and the DNA fragment of interest (gene) is inserted into a suitable, carefully selected vector, which might be a plasmid or a phage. The DNA of interest, placed into the vector, is then mixed with organisms whose cell walls have been treated in most cases chemically to become receptive and finally transformed. The transformation process is random and is not successful for every organism in one experiment. Therefore, an important step is screening for the transformants by using markers. Often the marker confers to the transformed organism an ability to survive under specific growth conditions and must be used with specially modified hosts that lack the specific trait. Selection of transformants is usually made through antibiotic resistance markers, supplementation of an auxothroph, and hydrolysis of a specific sugar. Having completed successfully the transformation process, the transformed organisms containing the DNA of interest are cultivated to provide a source of recombinant DNA. The recombinant DNA is then sequenced to verify whether it has the right sequence to make the enzyme. In the right case, a toxin-free organism is chosen with a demonstrated high capacity to produce foreign gene products and the gene of interest is cloned into this. Transformants are then cultured and the expressed enzyme is purified and analyzed using general protein chemistry methods to insure that it is identical to the native enzyme.

Recombinant DNA technology today is an efficient technology that makes a wide range of products for the food industry. Food-grade transformation systems are becoming available and many limitations of the past do not occur any more. However, commercialization of recombinant enzymes is very often limited by low yields. Improvement of yields to acceptable levels is a difficult, multidisciplinary task, which certainly limits the number of recombinant enzymes that reach the market. Current research focuses on molecular biology options to boost productivity, as well as on biochemical engineering strategies for suitable process designs. Genetically engineered enzymes are now available in the commerce and will continue to proliferate. Table 2 lists the most common food-processing enzymes extracted from genetically engineered bacteria and fungi and gives examples of their uses.

An example of a very successful application of genetic engineering in the area of food enzymes is that of recombinantly produced chymosin.[11] In the 1960s, the Food and Drug administration of the United Nations predicted a severe shortage of calf rennet, traditionally used in cheese making. It was anticipated that an increased demand for meat would lead to more calves being reared to maturity, so that less rennet would be available. In the early eighties, Gist-brocades started research on making calves' rennet by fermentation of a genetically modified microorganism. After an extensive evaluation of several possible hosts, it was decided to clone and express the DNA coding for calf stomach chymosin in an industrial strain of *Kluyveromyces lactis.* The natural habitat of *K. lactis* is milk and its products and the particular microorganism is known to be completely harmless and nontoxicogenic. The same strain was used for many years by Gist-brocades for the

Table 2. The most common food-processing enzymes made by genetically engineered bacteria and fungi. Examples of their uses

Enzyme	GMO	Examples of Uses
α-acetolactate decarboxylase	Bacteria	Removes bitterness from beer
α-amylase	Bacteria	Starch hydrolysis
Catalase	Fungi	Improves preservation in egg-based products
Chymosin	Bacteria/Fungi	Cheese production
Cyclodextrin-glucosyl transferase	Bacteria	Carbohydrate modifications
β-glucanase	Bacteria	Improves beer filtration
Hydrolases (various)	Bacteria	
Glucose isomerase	Bacteria	Carbohydrate modifications
Glucose oxidase	Fungi	Improves food preservation
Lactase	Fungi	Lactose hydrolysis
Lipases	Fungi	Oils and fats modifications
Pectinesterase	Fungi	Improves fruit juice clarity
Proteases (various)	Bacteria/Fungi	
Pullulanase	Bacteria	Carbohydrate modifications
Sulfhydryl oxidase	Fungi	Stabilizes the protein matrix of bakery products
Xylanase	Bacteria/Fungi	Enhances rising in bakery products

production of the food enzyme lactase which has been affirmed GRAS (generally regarded as safe) status by the US FDA. The strain proved to be a suitable host to express the DNA coding for calf preprochymosin. With the aid of the yeast α-factor leader as a signal sequence, the prochymosin is efficiently secreted into the medium. As *K. lactis* does not secrete appreciable amounts of endogenous proteins, the recovery of prochymosin is rather simple. The fermentation is followed by an acid step to autolyse prochymosin to active chymosin, the biomass is removed and no further steps are required to obtain a chymosin preparation, with a much higher purity than traditional calves' rennet. Recombinantly produced chymosins from *Escherichia coli* and *Aspergillus niger* are now available and their production by fermentation can provide a consistent supply of enzyme with properties analogous to those of the native enzyme.

Expression of the gene encoding chymosin in food-grade filamentous fungi, e.g., *Aspergillus niger*, has also been successful and led to a commercially useful product.[12] A system for controlled expression and secretion of bovine chymosin in *Aspergillus niger* has been reported by Cullen et al.[13] Protein expression in that case was regulated by starch induction and production processing was similar to that of glucoamylase.

The success of chymosin in the production of milk protein curd suitable for cheese making relies on the specificity of the proteolytic function. There are many proteases which cleave phenylalanine-methionine bonds but equally well cleave a variety of other peptide bonds. Nonspecific proteolysis results in a loss of curd volume and in the production of unpleasant flavours caused by proteolytic peptides. Chymosin, in the first steps of curd formation, removes a glycopeptide from χ-caseine by cleavage at a single phe-met peptide bond, destabilizing this way the casein micelles. The resulting para-χ-caseines aggregate in the presence of calcium ions to form insoluble complexes. Limited nonspecific proteolysis during the cheese maturation is desirable in this case because it contributes to flavour development. Chymosin obtained from recombinant organisms has been subjected to rigorous tests to ensure its purity. Biochemical, toxicological and microbiological tests have also been done. Companies producing the enzyme commercially by genetic engineering submitted data to FDA that demonstrated that recombinantly produced chymosin is chemically and kinetically identical to the native enzyme and that all safety and good manufacturing requirements are met. An additional advantage lies in the absence of contaminating pepsin. Through genetic engineering, the food industry and the consumer are the beneficiaries of a supply of consistently high quality product at a stable price.

Immobilized Enzymes

Enzymes, like other catalysts, are not consumed in the reaction they catalyze. In most applications however, they are used in a way similar to other chemicals and recovery steps are not included. Due to the usually high cost of the enzymes used in food processing it is advantageous to the process economics to have a system whereby the enzyme can be recovered and reused. A method to achieve these aims is enzyme immobilization. This involves the presentation of an enzyme in an insoluble form as a macromolecular matrix. There are several types of media available for the immobilization of enzymes, e.g., polyacrylamide, Nylon, γ-alumina, glass, cellulose, and several enzymes important to the food industry have been successfully immobilized, e.g., polygalacturonidase, lactase, urease, glucose oxidase, invertase.

The first publication on enzyme immobilization was by Nelson and Griffin in 1916[14] who reported that invertase had been adsorbed on to aluminum hydroxide while it retained its catalytic activity. Since then, numerous efforts have been devoted to the development of insoluble immobilized enzymes for various applications. There are several benefits of using immobilized enzymes rather than their soluble counterparts. Immobilized enzymes are more resistant to denaturation and proteolysis. Immobilized enzyme preparations can be recovered and recycled in continuous reactor systems or used over long periods in plug-flow reactors. Despite the increased costs of manufacturing immobilized enzyme preparations, a net positive cost benefit might result from the reuse and the modified properties they attain through immobilization. However, despite these apparent advantages, relatively little of the world's current enzyme-based industry utilizes immobilized enzyme technologies. Only one process, the conversion of glucose to fructose by glucose isomerase in the production of high fructose syrups, uses immobilized enzymes at a rate of more than 50 tons per annum.[15] Several other immobilized enzymes are used at much lower rates, e.g., β-galactosidase in the hydrolysis of milk lactose. Processes also exist in which the soluble enzyme option remains by far the preferred option. As an indication of the relative importance of immobilized versus soluble enzymes, it is significant that nearly 99% of the world's production of glucose is generated by soluble glucoamylase.

Immobilization techniques can be divided into physical and chemical methods. Physical methods include adsorption and entrapment. Chemical methods include covalent attachment to a matrix and the formation of a macromolecular matrix by cross-linking the enzyme. Figure 2 gives a diagrammatic representation of immobilization techniques. An enormous number of different types of matrices have been used in laboratory immobilization studies. Factors to be considered in the selection of an immobilization matrix include the cost, chemical resistance and physical properties of the matrix, the immobilization chemistry and the stability of the enzyme-matrix association and finally, the special requirements of the process.[15,16] The chemistry of covalent enzyme immobilization is highly complex. The successful preparation of an immobilized product requires an understanding of the nature of the reactive groups both on the matrix and the enzyme. For example, it is most important that the immobilization chemistry is selected to minimize the covalent modification of active site amino acid side chains. There are cases in which immobilization appears to have little or no effect on enzyme kinetics and other reaction characteristics. In other cases, immobilization can result in severe alterations to these important parameters. A characteristic example is the shift in the pH optimum to a lower value which is common with enzymes immobilized to positive net charge supports.[17] However, in some cases the resulting alterations in operational parameters of enzymes may become advantages for certain applications. Table 3 gives a summary of advantages and disadvantages of enzyme immobilization techniques.

Immobilization methods in which chemical bonds are formed involve the risk that essential functional groups in the active site of the enzyme will take part in the reactions resulting in loss of catalytic activity. To avoid this effect, immobilization is carried out in the presence of a suitable substrate that binds the active site in a way that protects it. Various other immobilization effects

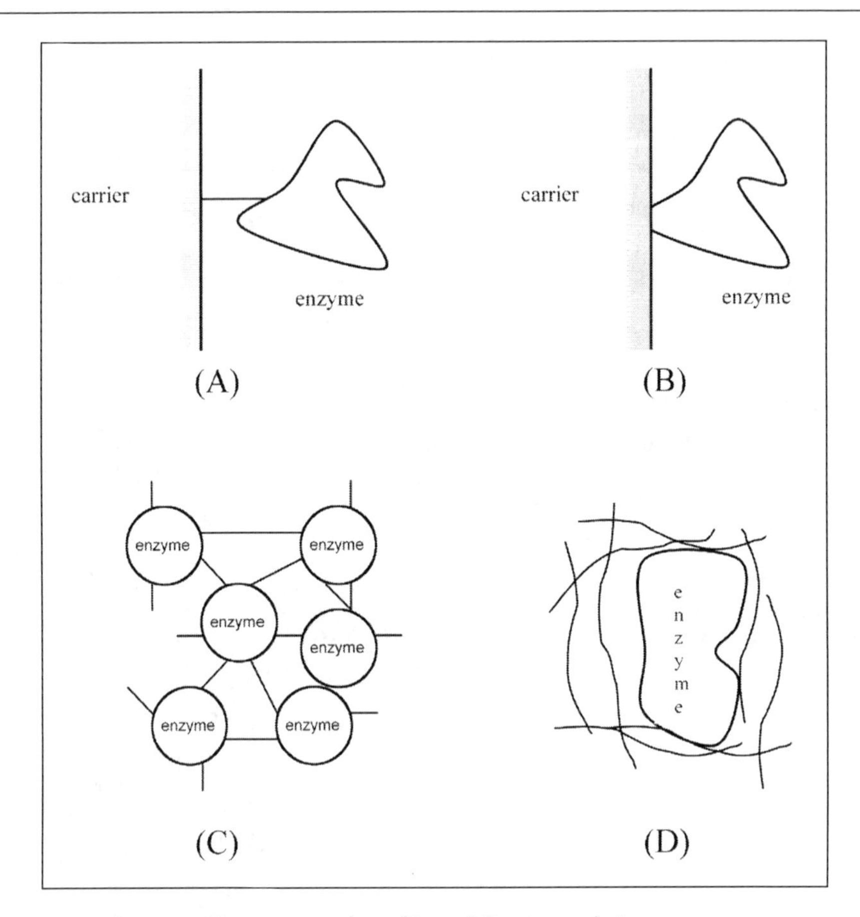

Figure 2. A diagrammatic representation of immobilization techniques.

include mass transfer effects, pH shifting, partitioning effects, which are important when enzymes are used in organic media, as well as several specific stabilization effects. For example, dissociation of oligomeric enzymes can be prevented by immobilization and proteolytic enzymes are stabilized by attachment to supports because their autolysis is prevented. Enzymes inside solid particles are protected from microbial degradation and proteolytic breakdown. The support can be designed to contain certain functional groups or additives, giving protection from inactivating reagents. For example, metallic oxides or catalase have been added to decompose hydrogen peroxide. The use of microscopic carriers in which enzymes are encapsulated, for example liposomes, offers this possibility and encapsulated enzymes are protected against adverse environmental conditions by additionally encapsulating additives such as antioxidants, chelating agents, buffers and possibly the substrate of the enzymes. While encapsulated, the enzyme is passive and latent within the food matrix. The design of the carrier determines where and when the enzyme is released and allowed to react with its substrate. The stability of the carrier can be controlled so that release occurs either early or late in the life of the food. The longevity of the liposomes can be varied by incorporating cholesterol[18] or α-tocopherol[19] so that there is flexibility in the release rate. The liposome technology has been applied in the production of hard cheeses,[20] for the prevention of cheese spoilage, and the antioxidant protection of polyunsaturated fats in various food products, such as salads and dressings.[21]

The literature information on immobilized enzymes is growing enormous, so does the list of commercially available water-insoluble enzymes and various immobilized systems. However,

Table 3. A summary of advantages and disadvantages of enzyme immobilization techniques

Method	Advantages	Disadvantages
Covalent attachment	Not affected by pH and substrate concentration	Possible modifications in active site. Expensive method
Adsorption	The simplest and less costly immobilization method. Enzyme remains unaffected. Support material can be retrieved	Leakage can occur if the conditions in the reaction mixture are drastically different from those used in adsorption. Enzyme subject to microbial or proteolytic enzyme attack
Covalent cross-linking	Enzyme strongly bound, thus unlikely to be lost	Method not effective for macromolecular substrates. Regeneration of carrier not possible. Possible loss of enzyme activity during preparation
Entrapment	No chemical modification of enzyme. Enzyme not subject to microbial or proteolytic action	Preparation difficult and often results in enzyme inactivation. Not suitable for macromolecular substrates. Diffusion effects affect transport or substrate to and product from the active site

enzyme immobilization is experiencing an important transition today. The recent review by Cao[22] summarizes the state of the art in enzyme immobilization methodologies and analyzes the trend of development with respect to rational design and combination of various immobilization methods or disciplines, with the aim of achieving the desired catalytic and/or noncatalytic functions. Combinatorial approaches are increasingly applied in the design of robust immobilized enzymes by rational combination of fundamental immobilization techniques (i.e., noncovalent adsorption, covalent binding, entrapment and encapsulation) or with relevant technologies. The objective is to solve specific problems that cannot be solved by one of these basic methods. Research efforts are focusing towards the design of immobilized enzymes with high volume activity by using for example carriers with high payload[23] or by encapsulation of carrier-free immobilized enzymes.[24] Another area is the design of immobilized enzymes with high stability which can be achieved through microenvironment engineering which is also applied for the design of immobilized enzymes with improved selectivity as well as improved noncatalytic functions.

Abzymes (Catalytic Antibodies)

Early in the 20th century, Emil Fisher suggested that enzyme action depends on the geometric structures of both the enzyme and the substrate and that they must fit like lock and key. Linus Pauling (1948) extended the lock-key model by suggesting that the structure of the active site of an enzyme is complementary to the transition state of the substrate-product. He suggested that an enzyme has a structure that is closely similar to antibodies, with the exception that it is not the surface configuration that is similar to the transition state but the active site region. These observations have given scientists reason to wonder whether antibodies made to bind a transition state model would catalyze that chemical reaction.

Early attempts to isolate antibodies with catalytic properties, after inoculation of animals with transition state analogues, were not successful.[25,26] The advent of monoclonal antibodies has been used extensively towards the production of catalytic antibodies and production of catalytic

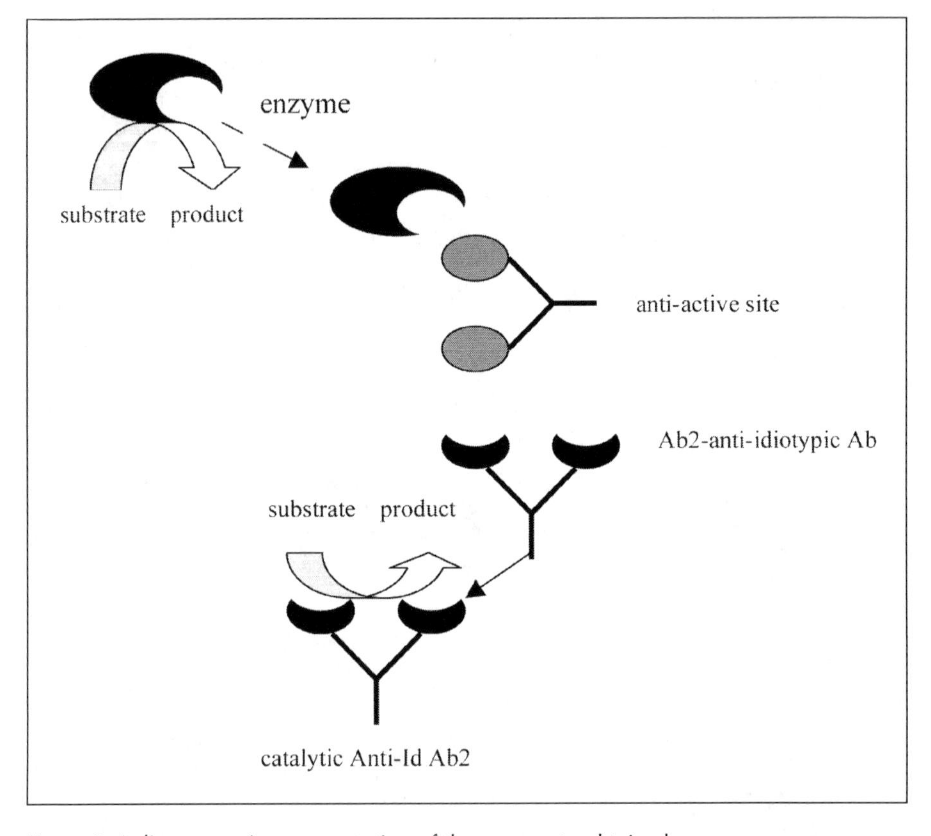

Figure 3. A diagrammatic representation of the process to obtain abzymes.

monoclonal antibodies was reported by several groups in 1985.[15] Richard Lerner's group at Scripps Research Institute generated antibodies to a series of tetrahedral phosphorus compounds that mimic the presumed transition state in the hydrolysis of a carboxylate ester. Lerner is credited with the original work that enabled monoclonal antibodies to be raised against the correct transition state analogues that would lead to enzyme-like reactions.[28] These catalytic antibodies exhibited properties of an enzyme, such as substrate specificity, saturation kinetics and competitive inhibition and were able to catalyze the reaction at a rate of 1,000 times greater than the uncatalyzed reaction. The catalytic antibody approach uses enzymatic mechanistic knowledge to provide new insights into the design of catalytic function and the use of such well-established studies avoids many of the questions regarding the structure-function correlation.[15] Most transition state analogues that have been synthesized so far are dependent on the use of a tetrahedral phosphorus derivative to mimic the acyl-transfer reaction. These antibodies have been shown to be able to catalyze reactions like the hydrolysis of several esters, hydrolysis of *p*-nitroanilide and the intra-molecular lactonization of a xydroxyester.[27,28] Figure 3 shows a diagrammatic representation of the process to obtain abzymes. The first antibody Ab1 is raised that recognizes the active site of an enzyme in a way that its combining site has structural features complementary to those of the enzyme. Against the Ab1 combining site, a second set of antibodies Ab2 is generated. These are the anti-idiotypic antibodies which may represent a structural internal image of the original enzymatic site and in some cases exhibit catalytic activity. Using this approach, antibodies with esterase and amidase efficient catalytic activities have been produced and characterized.

Various other approaches for the production of abzymes have been reported. One of them makes use of specificity of antibodies for molecules that are not transition state analogues, in which case the catalysis is carried out by an antibody cofactor combination.[29] Industry in general, including the food industry, finds the catalysis of reactions using cofactors uneconomic, mainly because of the high costs involved in the continual recycling.[30] However, research efforts are directed towards the design of stable antibody cofactor complexes that would not require recycling. The number of reactions catalyzed successfully by abzymes is growing and so far more than 100 abzymes differing in specificity and mode of action have been described. These reactions include hydrolydis of chemical bonds, stereospecific synthesis of compounds, as well as reactions of isomerization, decarboxylation, oxidation and reduction. At present several efforts are underway to determine structural and mechanistic features of abzymes. Efforts are also underway to improve the efficiency of abzymes—which is often low compared to their natural counterparts—by random mutagenesis and other molecular biology techniques. The first obtained abzymes were characterized by rather low efficiencies. In the mean time, improvements in hapten design, in the strategies used for immunization, in site-directed mutagenesis, as well as in methods developed for screening and selection, allowed to obtain catalysts with efficiencies sufficient for industrial applications.

Abzyme technology holds enough promise and large multinational companies are involved in the area and produce patents in the field. Abzymes are biological enzyme mimics. Constructed mimics (artificial enzymes) however are attracting an increasing interest today.[31] According to research reports, mimics can be constructed based on a nonpolypeptide chain backbone, such as with amino acid residues attached to nucleic acid backbone or to chemical polymer backbones available or constructed to order. The latter might be able to withstand very high temperatures ($>150°C$) without loss of activity, or a conformational change. Wiseman et al,[32] in their *Trends in Biotechnology* comment give a very informative insight into the approaches being currently used towards construction of novel biocatalytic enzymes by directed evolution. Mimics can compete with biological enzymes in applications where outstanding pH and thermal stability are paramount, even though this is at the expense of turnover number in substrate conversion to product. The artificial enzymes area is in its infancy but, obviously, is where the enzyme industry heading.

Conclusions

In food technology, enzymes are, in most cases, called to perform against adverse environmental conditions—limiting factors to their applications. At this point, it is obvious that biotechnology's advances open new roads that provide a gradual but steady expansion in new enzyme applications. Problems related to specificity, thermostability and other functional characters are being overcome and the possibility of carrying out reactions without the disadvantages of relatively delicate macromolecules, while yet retaining the catalytic rate enhancement and specificity characteristics of an enzyme, is no more characterized as overoptimistic. Genetic engineering and biochemical engineering applications in industrial enzymology have led to increases in supplies and reductions in costs of many industrial enzymes of paramount importance in food production. There is no doubt that revolutionary changes in molecular biology are going to open thousands of new uses of enzyme technology. Biotechnology is a highly skilled and specialized science. And, as Drs Jo Wegstein and Henry Heinsohn[33] comment "the food industry cannot rely on microbiologists to propose new applications. Insight is needed to identify opportunities in which novel enzymes might excel. Such insight requires knowledge of the benefits and willingness by food scientists to work jointly with enzyme suppliers to pioneer new applications. The opportunities are unlimited."

References

1. Winter G, Fersht AR, Wilkinson AJ et al. Redesigning enzyme structure by site-directed mutagenesis: Tyrosyl tRNA synthetase and ATP binding. Nature 1982; 299:756-758.
2. Charles AD, Gautier AE, Edge MD, Knowles JR. Targeted point mutation that creates a unique Eco RI site within the signal codons of the β-lactamase gene without altering enzyme secretion or processing. J Biol Chem 1982; 257:7930-7932.

3. Sigal IS, Harwood BG, Arentzen R. Thiol-β-lactamase: Replacement of the active-site serine of RTEM β-lactamase by a cysteine residue. Proc Natl Acad Sci USA 1982; 79:7157-7160.
4. Dalbaldie-McFarland G, Cohen LW et al. Oligonucleotide-directed mutagenesis as a general and powerful method for studies of protein function. Proc Natl Acad Sci USA 1982; 79:6409-6413.
5. Petrounia JP, Arnold FH. Designed evolution of enzymatic properties. Curr Opin Biotechnol 2000; 11:325-330.
6. Diaz JE, Howard BE, Neubauer MS. Exploring biochemistry and cellular biology with protein libraries. Curr Issues Mol Biol 2003; 5:129-145.
7. Bott R, Ultsch M, Wells J et al. Three dimentional structures of enzymes engineered to have altered specificities. Genencor Technical Literature 1988, South San Fransisco, USA.
8. Oxender DL, Graddis TJ. Protein enineering. In: Biotechnology-The Science and the Business. Moses V, Cape RE, eds. Harwood Academic Publishers, Switzerland 1994:153-166.
9. Wassman CD, Tam PY, Lathrop RH et al. Predicting oligonucleotide-directed mutagenesis failures in protein engineering. Nucleic Acids Res 2004; 32:6407-6413.
10. Lee C, Levitt M. Accurate prediction of the stability and activity effects of site-directed mutagenesis on a protein core. Nature 1991; 352:448-451.
11. van Dijck PWM. Chymosin and Phytase. Made by genetic engineering. J Biotechnol 1999; 67:77-80.
12. Genencore: Filing of a petition for affirmation of GRAS status (Petition No. GRASP 9G0352). Fed Reg 1999; 54:40910.
13. Cullen D, Gray GL, Wilson LJ et al. Controlled expression and secretion of bovine chymosin in Aspergillus nidulans. Bio/Technol 1987; 5:369-376.
14. Nelson JM, Griffin EG. Adsorption of invertase. J Am Chem Soc 1916; 38:1109-1115.
15. Cowan D. Industrial enzymes. In: Biotechnology-The Science and the Business. Moses V, Cape RE, eds. Harwood Academic Publishers, Switzerland 1994:311-340.
16. Bornscheuer UT. Immobilizing enzymes: How to create more suitable biocatalysts. Angew. Chem Int Ed Engl 2003; 42:3336-3337.
17. Goldstein L, Levin Y, Katchalski E. A water-insoluble polyanionic derivative of trypsin. II Effect of the polyelectrolyte carrier on the kinetic behavior of the bound trypsin. Biochemistry 1964; 3:1913-1919.
18. Kirby CJ, Clarke J, Gregoriadis G. The effect of the cholesterol content of small unilamellar liposomes on their stability in vivo and in vitro. Biochem J 1980; 186:591-598.
19. Halks-Miller M, Guo LSS, Hamilton RL. Tocopherol-phospholipid liposomes: Maximum content and stability to serum proteins. Lipids 1985; 20:195-200.
20. Kirby CJ, Gregoriadis G. Accelerated ripening of cheese using liposome-encapsulated enzymes. Int J Food Sci Technol 1987; 22:355-375.
21. Kirby CJ. Delivery systems for enzymes and functional food ingredients. Chemistry in Britain 1990; 26:847-850.
22. Cao L. Immobilised enzymes: Science or art? Curr Opin Chem Biol 2005; 9:217-226.
23. Blanco RM, Terreros P, Fernandez-Perez M et al. Functionalization of mesoporous silica for lipase immobilization characterization of the support and the catalysts. J Mol Cat, B Enzym 2004; 30:83-93.
24. Cao L, van Langen L, Sheldom RA. Immobilised enzymes: Carrier-bound or carrier free? Curr Orin Biotechnol 2003; 14:387-394.
25. Raso V, Stollar BD. The antibody-enzyme analogy. Characterization of antibodies to phophopyridoxyltyrosine derivatives. Biochemistry 1975; 14:584-599.
26. Summers R. Catalytic principles of enzyme chemistry. Ph.D. Thesis. 1983. Harvard University, USA.
27. Tramantano A, Janta KD, Lerner RA. Catalytic antibodies. Science 1986; 234:1566-1570.
28. Pollak SJ, Jacobs JW, Schultz PG. Selective chemical catalysis by an antibody. Science 1986; 234:1570-1573.
29. Shokat KM, Leumann CJ, Sugasawara R, Schultz PG. A new strategy for the generation of catalytic antibodies. Nature 1989; 338:269-271.
30. Goodenough PW. Food enzymes and the new technology. In: Enzymes in food processing. Tucker GA, Woods LFJ, eds. Glasgow: Blackie Academic & Professional, 1995:41-113.
31. Petsko GA. Design by necessity. Nature 2000; 403:606-607.
32. Wiseman A, Goldfarb PS, Woods L, Ridgway T. Novel biocatalytic enzymes by directed evolution. Trends in Biotechnol 2001:19:382.
33. Wegstein J, Heinsohn H. Modern methods in enzyme expression and design. In: Nagodawithana T, Reed G, eds. Enzymes in Food Processing. San Diego: Academic Press, 1993:71-101.

Chapter 4

Food Manufacture

Katalin Bélafi-Bakó*

Abstract

A summary on the role and functions of enzymes and enzyme mixtures in food processing technologies (dairy and meat industry, fish processing, baking, starch and sugar industry, production of beverages and fruit juices, and fat and oil processing), in biosynthesis of food additives (surfactants, flavors, functional foods and sweeteners) and in the treatment of wastes and coproducts is presented in this chapter.

Introduction

Usage of enzyme systems in the food industry is considered as ancient as mankind itself. Food preservation, wine fermentation, and bread or cheese making are just a few examples from the dawn of history man.[1-4] However, only over the past few decades have great development in applied enzymology provided both the biochemical background and the technological/engineering application in the particular field of food processing.[5,6] During the research and development work, it often turned out that not a single enzyme specie but a mixture of enzymes are acting, in many cases synergistically and simultaneously in food processes. Thus several various enzymatic reactions take place resulting in a complex biochemical process. In this chapter, application possibilities of enzymes and enzyme mixtures in food technologies, in the production of food additives, and in the treatment of coproducts and wastes are summarized.

Enzymes and Enzyme Mixtures in Food Processing

Dairy Industry: Milk and Cheese Production

Seventy percent of the world population is lactose-intolerant: they cannot digest milk sugar. Lactose in the milk can be hydrolyzed by the β-galactosidase enzyme (E.C. 3.2.1.23.) into glucose and galactose.[4] The process results in lactose-free milk; moreover, it is very useful in the production of flavored milk drinks by reducing the sucrose requirements. In yogurt and cheese manufacture the process accelerates acidification since lactose hydrolysis is the rate-limiting step of the process; it makes it possible to apply organisms that are not able to use lactose as a sole carbon source. This is because nonsweet lactose is converted into sweeter monosaccharides. The enzymatic method is extensively used in whey processing as well.

In cheese production the most important step is coagulation of milk which is carried out by the chymosin enzyme (E.C. 3.4.23.4.). Chymosin (rennin) is an aspartyl protease naturally found in the fourth stomach of the unweaned calf. Nowadays recombinant chymosin is used predominantly; microbial coagulants and other sources are considered only in a small segment of the market due to their limited availability.

The chymosin enzyme splits the polymer chain in κ-casein of milk between phenylalanine and leucine, resulting in clotting. The process is essential for cheese production, as it makes it

*Katalin Bélafi-Bakó—Research Institute of Chemical and Process Engineering, University of Pannonia, Egyetem u. 2., 8200 Veszprém, Hungary. Email: bako@mukki.richem.hu

Enzyme Mixtures and Complex Biosynthesis, edited by Sanjoy K. Bhattacharya.

possible to produce curd and whey. Curd is then further processed on to cheese, while whey—as a coproduct—can be treated and some of its valuable compounds can be recovered.

Several enzymes play an important role in cheese maturation where the main aim is to produce the typical flavor and texture of mature cheese.[7] Traditionally, the maturation process needed a prolonged period, but its acceleration was extremely desired by cheese producers due to economic considerations. As a result of an intensive search in "accelerated ripening technology", a list of probable and possible enzymes was compiled: which enzymes can be used—as an enzyme cocktail—in protein and fat hydrolysis (proteinases, chymosin and lipases), in production of volatile sulphur compounds (demethiolase and desulphurylase enzymes) which are especially important in cheeses such as Cheddar, and in production of other flavor compounds (peptidases and esterases). A key process is during ripening protein hydrolysis, where a combined action of chymosin and proteinases (mainly from bacterial origin)—as a first step—degrade casein into peptides. Many of the peptides obtained taste bitter or sour, therefore, peptidases must break them down into amino acids and aroma compounds to enhance the flavor properties.

Meat Industry

The conversion of muscle into meat is a complex biochemical process, that happens while the tissue passes through a stage of marked rigidity, termed rigor. The duration of rigor stiffness depends on the species, between 2-4 h and one day. Loss of rigor stiffness, i.e., the process of post-rigor muscle softening, is known as conditioning in the meat industry and is caused by the action of muscle proteolytic enzymes.[8,9] The various muscle proteinases and peptidases contribute significantly to the post-mortem changes. These enzymes can be classified according to the functional group at the active centre (e.g., aspartate-, cysteine- or serine-proteinase), or to the preferred pH range (neutral and acidic). It seems—in the light of the latest literature reports—that calpains (nonlysosomal, cysteine-proteinases, optimal pH 6.5-8.0) are the most important enzyme system in post-mortem conditioning.[10]

The eating quality of meat highly depends on its texture; tenderness seems to be the most important factor for customers. Therefore, exogenous enzymes are used to improve tenderness. During the tenderizing process, the aims are to disrupt myofibrillar proteins and—in case of connective-tissue toughness—to break collagen fibers. Proteinases from various (plant, animal, microbial or even recombinant) sources, such as papain, chymopapain, ficin, bromelain or collagenases are applied as tenderizing enzymes.

Fish Processing

Enzyme actions in fish processing are traditionally restricted to endogenous enzymes in e.g., production of fermented fish sauce (a popular product in Far East) by using the natural proteolytic enzymes. Recently new applications have been reported where enzymes help in deskinning fishes (tuna and skate), removal of cod liver membrane, or hydrolyzing the tissues that envelop individual salmon and trout toe eggs by various proteolytic enzymes.[11]

Baking Industry

In the baking industry three main groups of enzymes play an important role: starch degrading enzymes, proteases and pentosanases.[12] Starch-degrading enzymes include α- and β- amylases, diastase, glucoamylase; their actions result in hydrolysis of starch which improves and controls dough-handling properties and product qualities. Proteinases are applied for destruction of gluten-protein cohesiveness in the manufacture of wafers, cakes and crackers on one hand, and for the selective and controlled modification of proteins in bread making on the other hand.[13] Pentosanases or hemicellulases are used to treat the nonstarch polysaccharides of wheat flour.

Starch and Sugar Industries

In this field, several enzymes (Table 1) are applied for various purposes for the manufacture of different products from starch and/or starch derivatives. It can be seen that—beyond hydrolytic enzymes (amylases) for liquefaction and saccharification of starch—numerous nonhydrolytic en-

Table 1. Enzymes in starch and sugar industry

Enzyme	Substrate	Conversion	Product	Note
Amylases	Starch	Complete hydrolysis	Glucose	Liquefaction saccharification
Amylases	Starch	Partial hydrolysis	Maltodextrin	
β-amylase	Saltodextrin	Hydrolysis	Maltose	Special purposes
Glucose isomerase	Glucose	Isomerisation	High fructose corn syrup	Sweetener
Glucose oxidase	Glucose	Oxidation	Gluconic acid	E.g., desugaring eggs
Branching enzyme	Amylopectin	Transfer	Branched chain	
Cyclomaltodextrin glucanotransferase	Maltodextrin	Intramolecular transglycosylation	Cyclodextrins	Clathrate-like compounds

zymes are listed. Glucose oxidase is used to desugare eggs (glucose removal) to prevent browning during spray drying (egg powder production). Branching enzyme (α-1,4-glucan) is able to form branch points in amylopectin chains—similarly to glucan synthesis in living cells, thus providing a novel route for new food ingredients with retailored chemical structure from starch.[6] Cyclodextrins having 6-8 membered rings behave as clathrate-like compounds since their cavities inside are able to accommodate hydrophobic 'guest' molecules.[14] Thus, cyclodextrins can be applied for molecular encapsulation in food industry.

Production of Beverages and Fruit Juices

Coffee production includes harvesting, processing, roasting and storage. Processing can be carried out under wet or dry conditions. Enzymes can be applied in wet processing where the aim is to speed up removal of the mucilage from the freshly harvested green coffee beans. Usually pectinases are used as a crude preparation.[15]

Tea is produced from the leaves of *Camellia sinensis* and can be consumed as green tea (simply dried leaves of the bush) or as black tea, which is more popular due to the complex color and flavor formed by the so-called "tea fermentation" process.[16] During this process polyphenoloxidase (E.C. 1.10.3.1.) and peroxidase (E.C. 1.11.1.7.) enzymes convert the polyphenolic compounds (catechins) of tea into two major groups of pigments: thearubigins and theaflavins responsible for color. Regarding the flavor aspects of tea, lipoxygenase (E.C. 1.13.11.12.) activity produces the well-known "leaf aldehyde", hexenal aroma component. Other enzymes play minor role in the fermentation.

In fruit juice production, pectolytic enzymes are commonly used to reduce viscosity of pulp, increase yields and to improve liquefaction and clarification.[17,18] Pectinases are classified into three main groups (Table 2): pectinesterases, depolymerising hydrolytic enzymes and lyases. These enzymes act synergistically during degradation of pectin.

Another possible purpose in citrus juice processing is de-bittering by enzymes.[19] The major bitter compound in citrus species is glycosylated flavanone naringin. Naringinase preparation (from e.g., *Penicillium* or *Aspergillus* spp) containing α-rhamnosidase (E.C. 3.2.1.40.) and β-glucosidase (E.C. 3.2.1.21.) enzymes are able to break naringin (Fig. 1) to a less bitter pruning (containing naringinin and glucose units) while rhamnose is formed.

In the production of beer, two biochemical processes take place.[20] During the brewing process (malting), starch degradation is carried out by amylase enzymes into glucose, maltose and malto- triose, while in the fermentation process these saccharides are converted into ethanol.

Wine making procedure nowadays needs pectolytic enzymes which are used before fermentation.[2,18] In red wine production, pectinase enzymes are applied to (i) reduce the juice viscosity by hydrolyzing the pectin (similarly to fruit juice production); and (ii) enhance the extraction of pigments (i.e., anthocyanins) and other valuable compounds from the grape skin cell.

Table 2. Pectinase enzymes

Name	EC Number	Substrate	Mechanism
Pectinesterases: catalysing deesterification of the methoxyl group of pectin			
Poly-methyl-galacturonate-esterase	3.1.1.11.	pectin	random
Hydrolases: catalysing the hydrolytic cleavage of 1,4-glycosidic bonds			
Endo-polygalacturonase	3.2.1.15.	pectic acid	random
Exo-polygalacturonase 1	3.2.1.67.	pectic acid	terminal
Exo-polygalacturonase 2	3.2.1.82.	pectic acid	
Endo-polymethylgalacturonase		pectin	random
Exo-polymethylgalacturonase		pectin	terminal
Lyases: catalysing the cleavage of glycosidic bond by transelimination			
Endo-polygalacturonate-lyase	4.2.2.2.	pectic acid	random
Exo-polygalacturonate-lyase	4.2.2.9.	pectic acid	terminal
Endo-polymethylgalacturonate-lyase	4.2.2.10.	pectin	random
Endo-polymethylgalacturonate-lyase		pectin	terminal

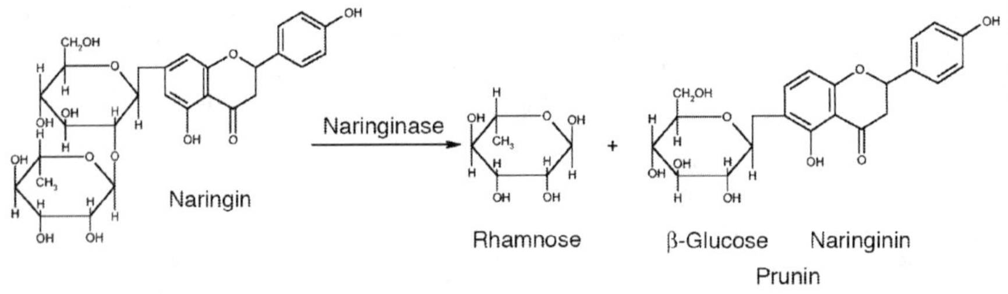

Figure 1. Enzymatic de-bittering of citrus juice.

Processing of Fats and Oils

Fats and oils (lipids) are chemically called triacyl-glycerols: fatty acid esters of glycerol. Lipase enzymes (E.C. 3.1.1.3.) are able to split the ester bond resulting in free fatty acids and glycerol.[21] This undesirable process unfortunately occurs in edible oils, increasing free-fatty-acid level. Lipases are responsible for hydrolysis of milk fat, as well, which can contribute to the cheese maturation process[22] on one hand, but it may be considered unfavorable causing butyric acid formation, on the other hand. Lipases can be applied, however, purposefully as process aids too. For example, lipases can be used in the manufacture of fatty acids from lipids, or in interesterification[23] for shifting the composition of common oils (e.g., palm oil) toward valuable cocoa butter (chocolate manufacture). Interesterification also takes place when polyunsaturated fatty acids derived from fish oils are incorporated into phospholipids by using lipase and phospholipase enzymes, improving nutritional quality.

Biosynthesis of Food Additives by Enzymes

Surfactants

Surface active agents are primarily used as detergents and emulsifiers in foods. Some of these compounds in high quality can be manufactured from renewable sources by enzymes. The polar part of these surfactants can consist of carbohydrates, while the apolar chains can be fatty acids, or their derivatives.[24] In this way sugars for example can be reacted with either fatty acids via an ester bond by lipase enzymes, or the derivatives via an amide functionality by other enzymes. These syntheses can be realized usually in non-aqueous media, such as an organic solvent.

Monoglycerides are another important group of surfactants in the food industry that are widely used as emulsifiers and can be produced by partial hydrolysis of triacyl-glycerols by lipase and fractionation.[25] The use of 1,3-regio-specific lipase is especially beneficial in providing 2-monoacyl-glycerol.

Flavors

Flavor substances include several chemically different compounds, such as aldehydes and (di)ketones, alcohols and acids, lactones, terpenes, isoprenoids and ionones.[2, 26] Among them esters are considered a very important group, with finding wide application possibilities in e.g., the beverage industry. Volatile, flavor ("fruit") esters can be manufactured if short chain alcohol compounds are reacted with short chain acids resulting in low molecular weight esters. The reaction is catalyzed by lipase enzyme (E.C. 3.1.1.3.) and can be carried out in an organic a solvent or solvent-free system under mild conditions.[27,28]

Functional Foods

This term was first introduced in Japan in the 1980s to describe food preparations which have a positive impact on human health. Functional foods are defined as processed foods containing or fortified with ingredients that have disease preventing and health enhancing benefits above their nutritive value and thus can be consumed as part of the daily diet. These food preparations can be divided into several categories: as dietary fibers, polyunsaturated ω-3 fatty acids, probiotic bacteria, phytochemicals and prebiotics.[29] Prebiotics have a significant beneficial effect to growing enterobacteria (*Bifidobacteria, Lactobacilus* species) which play an important role in human health. Moreover they form vitamins and other nutritive compounds necessary for the intestinal flora and prevent settling down of pathogenic microbes. These prebiotics include certain directly nondigestible oligosaccharides, like inulin derivatives,[30] fructo-oligosaccharides[31] and galacto-oligosaccharides among others. In the production of these compounds, enzymes can be applied either in partial hydrolysis reactions or in biosynthesis. During partial hydrolysis of malto- and pectic-oligosaccharides by amylase and pectinase enzymes, respectively, molecular weight distribution of the products should be controlled carefully, by methods such as membrane separation.[32] Fructo-oligosaccharides can be synthesized from saccharose by fructosyl-transferase enzymes (E.C. 2.4.1.99. and EC 2.4.1.100.), or similarly galacto-oligosaccharides from lactose by action of galactosyl-transferase in β-galactosidase.

Sweeteners

High fructose corn syrup (HFCS) is one of the major sweeteners in the beverage and baking industries.[5, 6] HFCS is produced from glucose syrup by using glucose-fructose isomerase enzyme (E.C. 5.3.1.5.), resulting in a much sweeter syrup, since fructose is about twice as sweet as glucose (Table 1).

Another natural, noncalorific sweetener, hesperidin, can be manufactured from citrus.[6] The original compound, a flavonone rutinoside, is virtually tasteless but its dihydrochalcone form (rhamnose is splitted from the molecule) becomes intensively sweet. The splitting can be carried out enzymatically using rhamnosidase.

Enzymes in the Treatment of Food Industrial Coproducts and Wastes

Dairy Industry—Whey

Whey is the coproduct of cheese production and is composed of lactose (70% of whey solids) and proteins (6%). In the processing of whey—after pasteurization—an ultrafiltration step is applied to concentrate the protein content. The rich-in-lactose permeate is either evaporated and dried lactose is manufactured (crystallization) or the lactose content is hydrolyzed by β-galactosidase enzyme resulting in a sweet syrup that is usable in the dairy, confectionery, baking and soft drink industries.[33]

Meat Processing—Bone, Feather, Miscellaneous

A significant amount of meat can be recovered from bones by enzymatic processes using papain, Neutrase and Alcalase proteinases from *Bacillus subtilis*. Neutrase e.g., provides a bland, low-fat protein powder, free of bitter peptides; Alcalase gives a lower grade product suitable for animal feed supplement and cleaned bones for high-grade gelatine production.[34]

Recently keratinases from various sources have been in the focus of many investigations.[35] Certain keratinases (from e.g., *Paecilomyces marquandii, Doratomyces microsporus*, or *Chryseobacterium sp*.) are able to degrade feather with an acceptable reaction rate resulting in a low biological value protein source.

Other types of animal wastes may be similarly good sources of protein hydrolysates. Atlantic cod viscera,[36] chicken intestinal,[37] shrimp head,[38] hake filleting waste,[39] or ram horn[40] are just a few examples of the wastes in animal processing available. Protein content of these materials can be hydrolyzed by various commercial protease enzyme preparations such as Alcalase, Neutrase or papain; or by different endo-and exopeptidases from *Aspergillus, Bacillus* species.

Fruit and Grape Juice Processing—Pressure Cake

In juice extraction by pressing, a cake is formed as a waste (mainly skin and seed) which may contain valuable compounds. In the case of grape, for example, seed oil can be extracted, but other pigments may also be worthwhile to recover. The major part of pectin—partially degraded in the pulp before pressing—remains in the cake as well, and its hydrolysis results in a valuable product, D-galacturonic acid. Derivatives of galacturonic acid can be applied in food industry as acidifying, tensioactive agents. The hydrolysis of pectin in the pressure cake can be completed by a mixture of pectinase enzymes.[17,18]

Processing of Crustaceans—Shells

Proteases are used to bioprocess an enormous waste mass, crustacean shells. Deproteinization of shrimp and crab shells has been carried out by Alcalase enzyme preparation resulting in a valuable chitin source.[41] Recently an enhanced method was presented, where protein hydrolyzate can also be recovered[42] increasing the process effectiveness.

Conclusions

Research and development on enzymes in the food industry is growing and expanding in the 21st century. Novel techniques with more active, specific and stabile enzyme preparations have been found for various purposes and new products have been developed. To follow the new findings, ideas, thoughts and results in this field, look at the scientific journals of the particular area such as the dairy industry regularly, study new patents, and scan the programs of special conferences (such as "Biocatalysis in the food and drink industries") in the interdisciplinary area of enzyme technology and food industry.

Acknowledgement

Part of the work was supported by GAK (MEMBRAN5) Project, grant No. OMFB-00971/2005.

References

1. Nagodawithana T, Reed G. Enzymes in Food Processing, 3rd ed. London: Academic Press, 1993.
2. Fox PF. Food Enzymology, London: Elsevier Applied Science, 1991.
3. Whitaker JR. Food Related Enzymes, Washington: ACS, 1974.
4. Dupuy P. Use of Enzymes in Food Technology, Paris: Technique et Documentation Lavoisier, 1982.
5. Birch CG, Balkebrough N, Parker KJ. Enzymes and Food Processing, London: Applied Science, 1981.
6. Tucker GA, Woods LFJ. Enzymes in Food Processing, 2nd ed. London: Blackie Academic and Professional, 1995.
7. Law BA, Mulholland F. Enzymology of lactococci in relation to flavour development from milk proteins. Int Dairy J 1995; 5:64-68.
8. Koohmaraie, M. Muscle proteinases and meat ageing. Meat Sci 1994; 36:93-104.

9. Penny IF. The enzymology of conditioning. In: Lawrie RA, ed. Developments in Meat Science. Barking: Applied Science, 1980:115-143.
10. Sorimachi H, Saido, TC, Suzuki, K. New era of calpain research: Discovery of tissue-specific calpains. FEBS Lett 1994; 343:1-5.
11. Wray T. Fish processing: New uses for enzymes. Food Manuf 1988; 63:48-49.
12. Kruger JE, Lineback D, Stauffer CE. Enzymes and their role in cereal technology. St. Paul: AACC, 1987.
13. Kczkowski J, Kurowska E, Moskal M. Value of cereal proteins. Proc Int Ass Cer Chem Symp Aminoacid Comp Biol Budapest 1983:391-394.
14. Huber O, Szejtli J, eds. Proc 4th Int Symp. Cyclodextrins. Dordrecht: Kluwer Academic Press, 1988.
15. Ehlers GM. Possible applications of enzymes in coffee processing. 9th Int Coll Chem Coffee. Paris: ASIC, 1980:267-271.
16. Sanderson GW, Coggon P. Use of enzymes in the manufacture of black tea. In: Ory R, St. Angelo AJ, ed. Enzymes in Food and Beverage Processing. Washington: ACS, 1977:12-26.
17. Pilnik W, Voragen AGJ. Pectic enzymes in fruit juice and vegetable juice manufacture. In: Reeds G, ed. Food and Science Technology, Enzymes in Food Processing. New York: Academic Press, 1993: 363-399.
18. Alkorta, I, Garbisu, C, Llama et al. Industrial applications of pectic enzymes: A review. Proc Biochem, 1997; 33:21-28.
19. Shaw PE. Removal of bitterness from citrus juices. In: Rousseff R, ed. Bitterness in food and beverages. Amsterdam: Elsevier, 1990:309-336.
20. Bamforth CW, Quain DE. Enzymes in brewing and distilling. In: Palmer GH, ed. Cereal Science and Technology. Aberdeen: Aberdeen University Press, 1989.
21. Adlerkreutz P. Enzyme catalysed lipid modification. Biotechnol Gen Eng Rev 1994; 12:231-247.
22. Arbige MV, Leung S. Isolation and characterization of a lipase for cheddar-cheese flavour development. JAOCS 1987; 64:645-647.
23. Macrae AR. Lipase catalysed interesterification of oils and fats. JAOCS 1983; 60:291-294.
24. Doren HA, Terpstra KR. Surfactants based on lactose. J Mater Chem 1995; 5:2153-2157.
25. Bellot JC, Choisnard L, Castillo E et al. Combining solvent engineering and thermodynamic modeling to enhance selectivity during monoglyceride synthesis by lipase-catalyzed esterification. Enz Microb Technol 2001; 28:362-369.
26. Berger RG. Aroma biotechnology. Berlin: Springer Verlag, 1995.
27. Gubicza L, Kabiri-Badr A, Keöves E, Bélafi-Bakó K. Large scale enzymatic production of natural flavour esters in organic solvent with continuous water removal. J Biotechnol 2000; 84:193-196.
28. Ehrenstein U, Kabasci S, Kümmel R et al. Entwicklung eines integrierten Verfahrens zur Herstellung natürlicher Aromaester. Chem Ing Techn 2003; 75:291-294.
29. Mizithrakis P, Tzia C. Concept of functional foods and future trends. Proc Biocat Food Drinks Ind London: SCI 2002:8-10.
30. Yun JW. Fructooligosaccharides—Occurrence, preparation, and application. Enz Microb Technol 1996; 19:107-117.
31. Cho YJ, Sinha J, Park JP et al. Production of inulooligosaccharides from inulin by a dual endoinulinase system. Enz Microb Technol 2001; 29:428-433.
32. Olano-Mamrtin E, Mountzouris KC, Gibson GR et al. Continuous production of pectic oligosaccharides in an enzyme membrane reactor. J Food Sci 2001; 66:966-971.
33. Kümmel R, Robert J. Application of membrane processes in food technologies. In: Bélafi-Bakó K, Gubicza L, Mulder M, eds. Integration of Membrane Processes into Bioconversions. New York: Kluwer Academic, 2000:143-163.
34. Sorenson NH, Rasmussen PB. Enzymic bone cleaning and scrap meat recovery. Proc 35th Int Cong Meat Sci Technol Copenhagen 1989:957-962.
35. Brandelli A, Riffel A. Production of an extracellular keratinase from Chryseobacterium sp. growing on raw feathers. Electr J Biotechnol 2005; 8:562-565
36. Aspmo SI et al. Enzymatic hydrolysis of Atlantic cod viscera. Proc Biochem 2005; 40:1957-1966.
37. Jamdar SN, Harikumar P. Autolytic degradation of chicken intestinal proteins. Bioresource Technol 2005; 96:1276-1284.
38. Ruttanapornvapeesakul Y et al. Effect of shrimp head protein hydrolysates on the state of water and denaturation of fish myofibrils during dehydration. Fisheries Sci 2005; 71:220-228.
39. Martone CB et al. Fishery by-product as a nutrient source for bacteria and archae growth media. Bioresource Technol 2005; 96:383-387.
40. Kurbanoglu EB, Algur OE. A new medium from ram horn hydrolysate for enumeration of aerobic bacteria. Turkish J Veter Animal Sci 2004; 28:343-350.
41. Oh YS et al. Protease produced by Pseudomonas aeruginosa K-187 and its application in the deproteinization of shrimp and crab shell wastes. Enzyme Microb Technol 2000; 27:3-10.
42. Synowiecki J, Al-Khateb NAAQ. The recovery of protein hydrolysate during enzymatic isolation of chitin from shrimp, Crangon crangon processing discards. Food Chem 2000; 68:147-152.

Chapter 5

Enzyme Applications in Leather Processing:
Current Scenario and Future Outlook

Palanisamy Thanikaivelan, Jonnalagadda Raghava Rao,* Balachandran Unni Nair and Thirumalachari Ramasami

Abstract

Global leather industry is undergoing paradigm shift from chemical to bio-based leather making to meet the growing environmental challenges. A short appraisal on the use of enzymes in current leather processing has been presented. Although enzymes find use in many areas of leather processing, enzyme-only dehairing and fibre opening presents a breakthrough approach. This methodology is being explored to avoid obnoxious sodium sulphide and to eliminate lime sludge. Enzyme applications for some unusual areas of leather making are also briefly outlined.

Introduction

An inevitable output from the production of almost all goods and services is pollution of some kind. Because of this the global environment is gradually degrading. Processing industries, which cause adverse changes to the immediate environment, are being challenged locally by the society. Leather industry is one among them. Leather and environment have been the two sides of a coin. In a way, leather making can be presented as an activity that helps in utilizing a potential waste.[1] The utilization of coproduct of the meat industry, namely hides/skins, is achieved by tanning. However, this is a simplistic view. A more detailed analysis of environmental consequences of tanning industry is critical. The industry has gained a negative image in the society due to its polluting nature. This is in spite of the leather industry having made traceable and visible impact on the socio-economics through both employment generation and export earnings.

Leather processing involves a series of steps (Fig. 1). They can be classified in four major groups viz., pretanning or beamhouse processes, tanning, post tanning and finishing as detailed by Ramasami and Prasad.[2] Pretanning processes aim at cleaning of hides/skins, tanning stabilizes the skin/hide matrix permanently and aesthetic values are added during post tanning and finishing operations. Various chemicals are employed in leather processing and varieties of materials are expelled.[3] The leather industry employs about 35-40 L of water per kg of hide processed.[2] With the present annual global processing capacity of 9×10^9 kg of hides and skins, it could be estimated that nearly 30 to 40×10^{10} L of wastewater is generated. This gives rise to two major problems for the leather industry, viz., the availability of good quality water and the need for treatment of such large quantities of effluent. The extent of pollution load emanating from the leather processing using conventional methods can be assessed from the emission factors for various operations (Table 1).[4] Nearly, 70% of the emission loads of biological oxygen demand (BOD), chemical oxygen demand

*Corresponding Author: Jonnalagadda Raghava Rao—Chemical Laboratory, Centre for Leather Apparels and Accessories Development, Central Leather Research Institute, Adyar, Chennai 600 020, India. Email: clrichem@mailcity.com

Enzyme Mixtures and Complex Biosynthesis, edited by Sanjoy K. Bhattacharya.

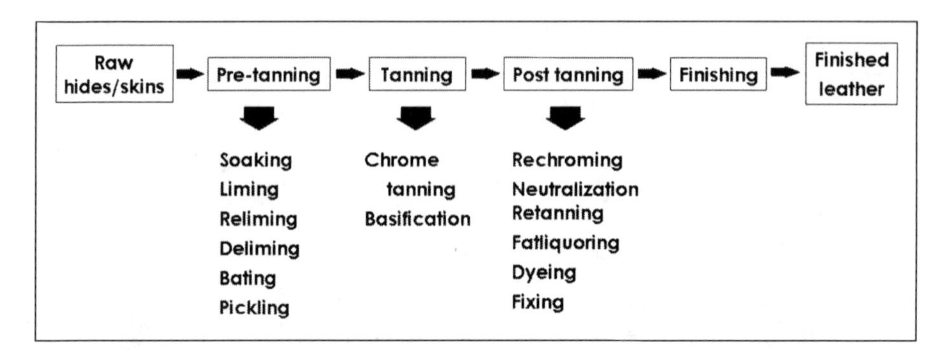

Figure 1. Process flow scheme of conventional leather processing.

(COD) and total dissolved solids (TDS) emanate from the pretanning processes.[5] Almost entire quantity of sulfide discharge is from dehairing process.[6] Chrome tanning activity is associated with large discharge of chromium and sulfate ions.[7]

Cleaner Leather Processing: Existing Enzyme Applications

Prevention or reduction of pollution at source through in-process control measures assumes greater significance now in leather industry due to the realization that end-of-pipe treatment alone is not enough to meet the stringent specifications laid down for the discharge of treated wastewater by pollution control authorities. The strategy for in-process control measures for pollution reduction should attempt to integrate cleaner process options with efficient water management practices as the volume of effluent has a direct influence on the cost of effluent treatment.[8] The reuse of spent liquor after the removal of the pollutants in suitable unit operation should be considered. The ideal strategy should be aimed at the zero or near-zero discharge of waste liquors.[9] The cleaner processing options recommended need to be cost-effective in order to be economically viable. The costing should take into consideration of the cost of the treatment of effluents in the absence of such options. The success of cleaner technologies depends on the following factors: (a) reduction of pollution in terms of quantity and quality, (b) tanners' benefits in terms of leather quality improvement and/or cost reduction, (c) reproducibility of the process, (d) cost effectiveness to be economically viable and (e) wide market opportunities.

Biotechnology, in its broad sense, has already been used in the tanning industry for a number of years now, since the inception of the use of enzymes. The utilization of enzymes in the tanning industry is possible in all stages of production, except perhaps the actual tanning. At the moment, the processes in which enzymes are already being used with a fair degree of success are: soaking (cleansing and rehydration), unhairing (removal of hair), liming (fibre opening), bating (removal of unwanted proteins) and, in part, degreasing (removal of fat). It should be mentioned that the use of enzymes in all these unit processes is not necessarily to achieve cleaner processing. Especially, in soaking and bating processes, enzymes are used to achieve desired action on certain noncollagenous proteins in preference to chemicals and not aimed to reduce the pollution load. Although several biotechnological options are available for handling effluents as well as proteinous solid wastes, the review focuses only on in-plant biotechnological leather processing.

Soaking

Soaking enzymes are mainly used to shorten the production time by disintegrating solidified noncollagenous proteins and fats that interpose themselves between the fibres and sometimes cover the external surfaces of the hide making contact between collagen fibres and water difficult.[10] Use of enzymes is not aimed for a specific action but rather a broad-spectrum of action in order to obtain both the solubilization and removal of the interfibrillar proteins for easy rehydration of the skin. Carbohydrases and proteases are the types of enzymes used in the soaking of hides. The

Table 1. Typical range of emission factors for conventional leather processing

Parameters	Soaking	Liming	Deliming	Pickling	Vegetable Tanning	Chrome Tanning	Dyeing and Fatliquoring	Composite (incl. washing)
BOD	6-24	15-40	1-6	0.2-0.7	3-18	0.3-1.6	1-4	30-120
COD	18-60	30-100	2.5-14	0.5-3	7.5-40	1-5	2-14	75-320
Total solids	200-500	90-200	4-20	17-70	12-60	30-120	4-20	450-1000
Dissolved solids	190-400	70-120	2.5-10	17-70	10-50	30-100	3-15	300-800
Suspended solids	15-60	15-80	1.5-8	0.5-3	2-10	1-5	0.6-2	60-160
Chloride as Cl^-	90-250	10-30	1-4	10-30	0.5-2.5	15-50	0.5-2	150-350
Total chromium	-	-	-	-	-	2-10	0.04-0.2	3-10

All values expressed in kg/t of hide or skin processed; These were obtained from the formula (concentration × volume of effluent)/t of leather processed.

optimum pH of operation can range from 4.0 to 10.5. The advantages of an enzyme soak include short wetting, loosening the scud (remnants of hair, epithelial tissue), initiation of the opening of the fiber structure, and, when used at an alkaline pH of less than 10.5, production of a product with less wrinkled grain.[11] The major disadvantage of their use is the added expense.

Grimm reported a soaking process using proteolytic enzymes and carbohydrases at pH 5.5-10.0.[12] Trabitzsch accounted on the possibilities of using enzymes in soaking.[13] Toshev and Esaulenko wet back nonsalted, preserved sheepskins using the enzymes from *A. oryzae* in 4 to 5 hrs.[14] Botev and coworkers used 0.5-0.6% of bacterial α-amylase for soaking dried wool lambskins and found that it resulted in stronger amylolytic activity, much weaker proteolytic activity and no lipolytic activity.[15] Pfleiderer accounted on the advantages of using mold proteases at pH 5 or lower for soaking, unhairing and bating.[16] Monsheimer and Pfleiderer patented a process for soaking with pepsin and papain at a pH of 3.0-4.5 and another process using alkaline proteinase at 28°C for 4 hours on salted calfskin.[17,18] They also suggested a recipe for soaking hides, skins and fur skins based on proteolytic enzyme at pH 10.5.[19] After 4 hours the hides were uniformly soaked and showed no evidence of hair slip.

Liming and Reliming

The conventional liming process employs lime and sulfide in high proportions.[20] These materials form a source of pollution in the spent lime liquors. Lime being a poorly soluble alkali, there is an advantage from limited availability of dissolved alkali. There is also disadvantage of generation of large quantum of solid wastes. Although sulfide is a toxicant, it is the prime depilant in the unhairing process. Reduction of sulfide at source is now possible using enzyme assisted processes.[21] Enzymes used in unhairing are generally of the proteolytic type that catalyze the breakdown of noncollageneous proteins.[22] Their origin can be animal, such as bovine or porcine pancreas, bacterial, fungal and plant.[22,23] Röhm described the first successful enzymatic unhairing, Arazym process,[24] as stated by Green.[25] Enzyme assisted unhairing (using enzymes, generally based on protease, along with small amounts of sulfide and lime applied as paint on flesh side) causes loosening of hair by selective breakdown of cementing substances and presents a hair saving approach.[26-31] Modifications were made by pretreating the hides or skins either with bases or disulfide cleaving compounds for accelerating the process.[32,33] Erhazym process was reported by Pfleiderer, which employed proteinases along with small quantities of sodium sulfide and sodium sulfhydrate.[34] Fibre structure of the enzyme unhaired skin was compact, however, lacked plumping.[35] Some researchers reported on the possibility of solubilization of collagen leading to damage in fibre structure.[36,37] Simoncini and Meduri assessed the activity of various enzymes for their unhairing activity and reported that Keratinase, Nercozyme 150, Arazym, Pronase, Napase, Bioprase and Protease 306 were highly active while pepsin, β-chymotrypsin, elastase, trypsin and α-chymotrypsin were either inactive or showed poor activity.[38] Andrews and Dempsey[39] and Hannigan et al[40] suggested additional reliming process to give a better product. Yates attempted to uncover the mechanism of enzymatic depilation process.[41]

Enzyme assisted processes for the removal of hair is associated with both merits and demerits.[22,42] The recognized merits are (a) higher area recovery,[43,44] (b) smoother grain, (c) better in-plant ecology and (d) higher efficiency of wastewater treatment systems. The demerits perceived are (a) fear of possible damage of leather making substance, (b) inadequate fibre opening, (c) flatter grain and substance and (d) higher chemical costs. However, it is possible to overcome the demerits through proper process optimization and control by employing a combination of enzyme with sulfide (Technology plan for environmental sustainability of the tanning sector in Tamilnadu, Status report submitted to AISHTMA by CLRI, Chennai, 1997).

Bating

In bating, the hides and skins are treated with proteolytic enzymes to remove certain undesirable noncollagenous proteins. McLaughlin and coworkers established that the bating enzymes bring about physico-chemical changes in the skin.[45] They concluded that the most vital function of the enzymes in the bate is the removal of the coagulable or coagulated protein of the skin. Wilson and Merrill demonstrated that the methods described for measuring enzyme activity upon collagen, elastin, and keratose are reliable guides for determining what a pancreatin will do in bating to the constituents of the skin.[46] Quarck reviewed and discussed about the theories and mechanism of bating.[47] He suggested that the mechanism of bating is substantially more complicated than most chemists are willing to admit. The effect of bating on individual properties of the leather such as break, temper, drawn grain etc, was studied by Stubbings.[48] Gustavson developed a new method for investigating the nature of the bating process, particularly the alterations of the hide substance by the proteolytic enzymes of the bate.[49] Moore demonstrated that the pancreatic enzyme based bate is superior to the fungal and bacterial enzyme bate, although the difference is not large; at higher temperatures, bating enzymes are more efficient than at lower temperatures.[50] Uehara and coworkers examined the changes in grain structure and physical properties of pig skin during bating process.[51]

Degreasing

Degreasing is another stage of the tanning process in which the use of enzymes is foreseeable and feasible. The quantities of natural fats are very high, especially in some types of sheepskins, wherein they can represent 20-30% of the total weight of the skins.[52] The normal fat removal methods involve the use of anionic or non-ionic surface active agents or solvents. The surface active agents make it possible to carry out the process in the aqueous phase but the results are not always satisfactory. Solvents are more effective but necessitate a series of preliminary operations that increase the costs and prolong the processing times apart from the fact that they cause severe ecological problems. In principle, the degreasing process requires three consecutive stages namely breakdown of the proteic membrane of the fat containing sac, removal of the fat and emulsification of the removed fat in water or solubilization in solvent. If one of these phases is carried out incorrectly or inadequately then, the whole degreasing process will be inefficient. An enzymatic preparation should therefore have a triple action comprising of proteolytic, lipolytic and emulsification, for it to be an effective degreasant.

Trabitzsch evaluated the potential of lipases in degreasing skins.[13] Papp et al patented a process in which animal skins were treated with lipase and amylase in the presence of deoxycholic acid catalyst to remove lipids and noncollagenous proteins.[53] Jareckas employed a combination of two enzymes, Protosubtilin G2X and Lipavamorin G3X, for 18 hours at 39-40°C at a pH of 7.5 and succeeded in unhairing and degreasing pigskin.[54] Subsequently, Jareckas and Shestakova substituted Protofradin G3X for the Protosubtilin G2X, and found that this process improved the texture of pigskin, removed fat, mucosaccharides and substances other than collagen.[55] It was found that the structure became more open, retained more fat during fatliquoring, and produced more elastic and fuller leather than pigskin unhaired by conventional method. Yeshodha and coworkers used a fungal lipase at a pH of 3.2-3.6 at 37°C for one hour and successfully degreased wool sheepskins.[56] A series of papers were published based on degreasing of sheepskins with acid lipase from *Rhizopus nodosus*.[57-59] The enzyme hydrolyzed positions 1 and 3 of the triglycerides indicating the specificity for the hydrolysis of primary esters. Zhang and Zhang used alkaline lipase in combination with the proteinase and pancreatin in softening pig skin to improve degreasing effect.[60] Pfleiderer suggested that a combination of enzyme compatible surfactants with the enzyme has a synergistic effect in both soaking and bating and thereby ensuring optimum wet degreasing of rawhides, pelts and wet blues.[61] Pfleiderer et al degreased the hides by bating with proteolytic enzymes in the presence of surfactants.[62] In spite of these technological advancements, Taeger and Pfleiderer reported that enzymatic degreasing with lipases has not been recommended due to the cost and technical reasons.[63] It appears that a combination of enzymes might be necessary not only to breakdown the grease but also to release the breakdown products from the hide.

Progress in the Recent Past

In the past two decades, significant improvements in the production and application of enzymes in various leather processing steps have occurred.[64-66] One such area is the use of enzymes for the removal of dung from hides during soaking process. Dung, which contains cellulose, hemicellulose and lignin up to 80%, poses severe problems for tanners in the production of high quality leathers. It has been demonstrated that mixture of enzymes such as cellulase, xylanase and laccase, were able to remove dung from hide pieces under the laboratory conditions.[67] Evolution of recombinant DNA (deoxyribo nucleic acid) technology and protein engineering has led to the development of a vast range of proteases for application at soaking, unhairing and bating processes.[64,68-71] Recently, it has been shown that a neutral protease, dispase, is able to loosen the hair without damaging the fibrous collagen.[72] In a similar manner, efforts have also been made to design an ideal enzyme unhairing, in which enzymatic activity is oriented towards the epidermis thereby their action on collagen is minimized.[73,74] Attempts have been made to develop alkaline protease, from *Alcaligenes faecalis, Aspergillus tamarii* and *Streptomyces griseus*, capable of loosening the hair without any chemical assistance.[75-77] An alkaline protease has been produced from *Rhizopus oryzae* through solid state fermentation, which dehair the skins completely.[78] Commercial exploitation of such enzymes would depend on the cost-effective production and secured method of application at industrial level. Use of enzymes has been suggested for dehairing after shaving the hair mechanically, though the process has a potential to damage the grain.[79,80] Recent studies show that some bacterial cultures have keratinolytic activity, which could be used for depilation.[81-84] However, the effect of degraded keratin compounds on the environment needs to be studied before attempting commercial utilization. Lime-free fibre opening methods have been attempted scarcely. These include the use of strong alkalis,[85] *Lactobacillus* based enzymes,[86] lyotropic agents and bating enzymes.[43] However, these methods have limited applications owing to inappropriate opening up of fibre bundles resulting in poor leather qualities.

Scale-up studies of the production of proteases for bating, and possible use of local materials such as cow pancreas for bating have also been carried out.[87-89] The new development in this field lies in the use of enzymatic products capable of effecting the bating action even at acidic pH. Acid protease produced by *Aspergillus usamii* has been shown to be an ideal bating agent for sheep-pelt.[90] The study also showed that acid protease was more effective than neutral protease as a sheep-pelt bating agent. Bating with the halophile protease in 4M NaCl yields leathers with satisfactory physical characteristics.[91] Studies to find the efficiency of degreasing using enzymes reveal that pickling with acid proteases and lipases increases the effectiveness of degreasing while neutral lipases increase the efficacy during degreasing at neutral pH.[92] A combination of enzymes might be necessary not only to breakdown the grease but also to release the breakdown products from the hide. A study based on a commercial lipase produced from *Rhizopus arrhizus* shows that the lipolytic activity of the enzyme was maximum at pH 8.0 and the enzyme can work together with alkaline protease in the presence of a non-ionic surfactant.[93] Evidence for hydrolyzing ovine storage lipid using lipid hydrolases has been shown in an attempt to understand the mechanism of degreasing.[94]

Although studies on the use of enzymes for various stages of leather processing are numerous, commercial production and application in the leather industry is very limited.[95] Because a considerable portion of skin/hide is noncollagenic and hence must be removed during the beamhouse processing, leather production must yield significant quantities of organic wastes. A big contributor to this waste is degraded keratin. It is estimated that about 64% of total COD and 56% of the total TSS (total suspended solids) are produced from beamhouse processes.[96] However, only very few enzyme products, capable of saving the hair, are available in the market for dehairing application from various multinational companies. Demand for such enzymes among the tanners is also low due to the risk of collagen degradation. However, it is possible to overcome this limitation through proper process optimization and control by using a combination of enzyme with sulfide.[97] Redesigning the conventional production processes using biotechnological methods has been successfully undertaken by the leather industry in the last decade (Technology plan for environmental sustainability of the tanning sector in Tamilnadu, Status report submitted to All India Skin, Hide, Tanners and Merchants Association (AISHTMA) by CLRI, Chennai, 1997). Since the enzymatic bating process has become indispensable over the years, numerous commercial products are being used in the tannery. Although a few commercial degreasing enzyme products are available, their use is negligible among the tanners from developing countries. New avenues in the use of enzymes have been found, especially at tanning and post tanning stages. Protein cross-linking enzymes such as transglutaminases have been used as alternative tanning agents, however with no success.[98] Pepsin extracted from bovine or porcine stomach mucous, which can be added at pickling or after chrome tanning, offers improved yield and soft leathers at low cost.[99] Such enzyme treatments lead to area increase of up to 4-5% at industrial level trials.[100] Possible improvement in the penetration of dye has also been claimed. A combination of acid lipase and mild acid protease has been reported to clean the surface of chrome tanned stock of grease, scud and other stains and provides uniformly colored leathers.[101]

Recent Trends in Enzyme Based Leather Processing

Enzyme-Assisted and Enzyme-Only Dehairing

An enzyme-assisted dehairing process using reduced amount of sodium sulfide and commercial enzyme based on bacterial alkaline protease has been developed for cowhides.[102] The process does not require lime for dehairing thereby ensuring not only 100% elimination of lime sludge formation, but also complete removal of hair at pH 8.0 (Table 2). The results indicate that sulfide concentration of 0.5% and enzyme concentration of 1% (on soaked weight) is required for complete removal of hair. Out of the several application methods employed, painting on the grain side of the cow sides resulted in complete dehairing. It is shown that the developed dehairing process has an environmental benefit in reducing the COD and TS loads by 45 and 13%, respectively.[102] Lime free enzyme based dehairing methods have been standardized for goat and sheep skins. Although grain side application for hides is feasible, it is difficult for skins due to denser and lengthier hair. Various studies indicate that drumming with enzymes in a low float (water 10%) provides better results. It has been observed that 1% enzyme without any sodium sulfide results in complete removal of hair in goatskins.[103] Whereas, for sheepskins, a combination of 1% enzyme and 0.25% sodium sulfide is needed to achieve complete dehairing.[104] This could be due to the presence of natural fat in sheepskins, which hinders the penetration of enzyme into the skin matrix and subsequent action on the hair bulb. Now, it has been possible to eliminate sodium sulfide by the use of sodium metasilicate in the enzyme assisted dehairing process for cowhides (Table 2).[105] It has been shown that silicate enhances the activity of the enzyme thereby enables complete elimination of sodium sulfide as well as lime.

Enzymatic Fibre Opening

An enzyme based fibre opening process has been established for cowhides, goat and sheepskins.[104,106,107] Enzymes such as protease and α-amylase have been used to remove hair and protein-carbohydrate conjugates, respectively (Table 2). Scanning electron microscopic analysis reveal that the samples obtained by lime and enzyme based fibre opening, followed by chrome tanning process exhibited similar fibre opening. This methodology provides significant reductions in the pollution loads as well as dry sludge formation. This is one of its kinds in providing a paradigm shift from chemical to bioprocessing in the conversion of skins into leather. The process is proven to be technically as well as economically feasible. Recently, an attempt has been made to integrate the enzyme based dehairing and enzymatic fibre opening processes (Table 2).[108] The developed single-step process enables dehairing and fibre opening concurrently thereby offers reductions in time, labor and other utilities.

Table 2. Timeline for the development of enzyme based dehairing and fibre opening processes

Process	Development	Chemicals/Enzymes	Timeline
Dehairing	Enzyme assisted dehairing[26-31]	Lime + Na_2S + Protease	1950 - present
	Enzyme assisted lime-free dehairing	Na_2S + Protease[102,104],	2000
		Silicate + Protease[105]	2005
	Enzyme-only dehairing	Protease[103]	2003
Fibre opening	Enzyme based fibre opening	α-amylase[104,106,107]	2002
Integrated enzyme based dehairing and fibre opening		Protease + α-amylase[108]	2006

Enzyme-Driven Tanning Process

An enzyme-driven tanning process towards clear leather processing has been developed for hides and skins.[104,107,109-114] The bio-driven process provides leathers of similar properties to that of normal processed leathers. In addition, a reduction in chemical consumption by 98% offers an eco-benign enzyme based tanning process. The COD and TS emission loads are significantly reduced, especially for processing cow hides[109] and sheepskins[104] (Table 3). The reductions in the emission loads of COD and TS for the enzyme mediated three step tanning process are 71-84% and 80-84%, respectively compared to the conventional leather processing. This methodology also provides economic benefits in terms of water, power consumption and area yield.

Conclusions

The conventional leather process leads to the discharge of large amounts of pollutants due to enormous chemical usage. Currently, leather processing is undergoing transition towards meeting environmental challenges. A paradigm shift from chemical to bioprocessing is gaining importance. A comprehensive review on the use of enzymes in leather processing has been made. In the pretanning processes, proteolytic enzymes or mixture of enzymes are used to clean the skin. Enzyme-only dehairing and fibre opening methodologies to replace the currently employed chemicals are being looked for achieving cleaner leather processing. Modern biotechnology now offers solutions to overcome the difficulties in achieving substrate specificity as well as economic feasibility. Future avenues in the use of enzymes in leather processing have also been traced and their technical features and practicability analyzed.

Table 3. Comparison of pollution loads for conventional (C) and enzyme-driven (E) leather processing[a]

	Emission Load (kg/t of raw material[b] processed)		% Reduction in Comparison to Conventional Process	
Process	COD	TS	COD	TS
C_{cow}	31	218	-	-
E_{cow}	9	44	71	80
C_{sheep}	49	222	-	-
E_{sheep}	8	36	84	84

[a]Composite liquors were collected from all the processing steps excluding soaking, and analyzed for COD and TS and multiplied with the volume of effluent. [b]Weight of skins/hides before soaking.

References

1. Germann HP. The ecology of leather production—Present state and development trends. In: Science and Technology for Leather Into the Next Millennium. McGraw-Hill Publishing Company, 1999:283.
2. Ramasami T, Prasad BGS. Environmental aspects of leather processing. Proceedings of the LEXPO XV. Calcutta, 1991:43-71.
3. Rao JR, Ramasami T. Waste management in leather processing: A case of chromium. In: International Conference on Industrial Pollution and Control Technologies (ICIPACT-97). Hyderabad, 1997.
4. Ramasami T, Sreeram KJ, Gayatri R. Emerging leather processing strategies for waste minimization. In: Unit 1. Bac kground information and cleaner technologies in raw material preservation and in the beamhouse processes. UNIDO, 1998:183-97.
5. Ramasami T, Rao JR, Chandrababu NK et al. Beamhouse—and tanning operations: Process chemistry revisited. J Soc Leather Technol Chem 1999; 83:39-45.
6. Marsal A, Cot J, Boza E et al. Oxidizing unhairing process with hair recovery. Part I. Experiments on the prior hair immunization. J Soc Leather Technol Chem 1999; 83:310-15.
7. Rao JR, Nair BU, Ramasami T. Isolation and characterization of low affinity chromium(III) complex in chrome tanning solutions. J Soc Leather Technol Chem 1997; 81:234-38.
8. Parthasarathy K. Water management in tanneries. In: International conference on water management—Water 95. Madras, 1995.
9. Sykes R. Zero emissions-pattern for the future? Leather 1997; 199(2):35-38.
10. Simoncini A. Biotechnology in the tanning industry. J Am Leather Chem Ass 1987; 82:226-41.
11. Pfleiderer E. Auxiliary agents for the bovine tannery beamhouse. Leather 1985; 187(2):14-18.
12. Grimm O. Verfahren zur Herstellung gerbfertiger Blößen. German Patent DE1,230,169, 1966.
13. Trabitzsch H. Concerning the possibilities of employing enzymes in the beamhouse. J Soc Leather Technol Chem 1966; 50:382-89.
14. Toshev T, Esaulenko L. Kozh-Obuv Prom 1972; 13(4):12.
15. Botev I, Esaulenko L, Toshev T. Kozh-Obuv Prom 1976; 17(2):13.
16. Pfleiderer E. Zur anwendung proteolytischer enzyme in sauren pH-Bereich für weiche, enthaarung und beize. Leder 1968; 19:301-3.
17. Monsheimer R, Pfleiderer E. Verfahren zur enzymatischen Behandlung von Häuten und Fellen. German Patent DE1,800,891, 1970.
18. Monsheimer R, Pfleiderer E. Verfahren zum Weichen von Fellen oder Häuten. German Patent DE2,059,453, 1972.
19. Monsheimer R, Pfleiderer E. Verfahren zum Weichen von Fellen und Häuten. German Patent DE2,944,461, 1981.
20. Money CA. Unhairing and dewooling-requirements for quality and the environment. J Soc Leather Technol Chem 1996; 80:175-86.
21. Alexander KTW. Enzymes in the tannery—Catalysts for progress? J Am Leather Chem Ass 1988; 83:287-316.
22. Taylor MM, Bailey DG, Feairheller SH. A review of the uses of enzymes in the tannery. J Am Leather Chem Ass 1987; 82:153-64.
23. Adewoye RO, Lollar RM. Use of pulped fruit of Adenopus breviflorus (tagiri) as an unhairing agent and characterization of the enzymes extracted from the fruit. J Am Leather Chem Ass 1984; 79:446-61.
24. Röhm O. Verfahren zum Beizen von Häuten. German Patent DE200,519, 1910.
25. Green GH. Unhairing by means of enzymes. J Soc Leather Technol Chem 1952; 36:127-34.
26. Bose SM. Enzymatic unhairing. Leather Sci 1955; 2:140-44.
27. Pilawski S, Felicjaniak B. Die einwirkung von pankreasenzymen auf schweinshaute im enthaarungsprozeß. Das Leder 1976; 27:102-09.
28. Jones HW, Cordon TC, Windus W. Light leather from enzyme unhaired hides. J Am Leather Chem Ass 1968; 63:480-85.
29. Puvanakrishnan R, Dhar SC. Recent advances in the enzymatic depilation of hides and skins. Leather Sci 1986; 33:177-91.
30. Brady D, Duncan JR, Russell AE. A model for proteolytic depilation of skins. J Am Leather Chem Ass 1990; 85:334-42.
31. Hetzel LV, Somerville IC. Unhairing calfskins and side leathers by an enzymatic process. J Am Leather Chem Ass 1968; 63:90-107.
32. Simoncini A, Pezzo LD, Gelsomino M. L'epilage enzymatique des peaux de bovins pour la production de cuir a semelle. Rev Tech Ind Cuir 1966; 58:115-35.
33. Yates JR. Studies in depilation. V. Investigation of the relative depilatory efficiency of a number of commercially available enzyme systems. J Am Leather Chem Ass 1968; 63:464-73.

34. Pfleiderer E. Praxiserfahrungen mit dem enzymatischen Einstufenverfahren. Leder-und Hautemarkt 1977; 29:275-78.
35. Andrews RS, Dempsey M. The changes in fibre structure of ox hide on enzyme unhairing and tanning as sole leather. J Soc Leather Technol Chem 1966; 50:209-18.
36. Urbaniak MA. The influence of some enzyme preparations on components of skin Part II—Collagenous proteins and skin tissue. J Soc Leather Technol Chem 1973; 57:39-43.
37. Yates JR. Technical note on the practical application of novo dewooling enzyme No. 1. J Soc Leather Technol Chem 1973; 57:44-45.
38. Simoncini A, Meduri A. Unhairing activity of proteolytic enzymes. Cuoio Pelli Mat Concianti 1968; 44:347-56.
39. Andrews RS, Dempsey M. Some investigations into methods of unhairing II. Experiments on enzyme unhairing. J Soc Leather Technol Chem 1967; 51:247-58.
40. Hannigan MV, Happich ML, Jones HW et al. Sole leather from enzyme-unhaired hides. J Am Leather Chem Ass 1968; 63:522-29.
41. Yates JR. Studies in depilation. IX. Effect of skin thickness and diffusion on the rate of depilation of sheepskins. J Am Leather Chem Ass 1969; 64:71-81.
42. Feigel T. Use of enzymes in the beamhouse—possibilities and limitations. World Leather 1998; 11(3):54-59.
43. Thanikaivelan P, Rao JR, Nair BU. Development of a leather processing method in narrow pH profile. Part 2. Standardisation of tanning process. J Soc Leather Technol Chem 2001; 85:106-15.
44. Shrewsbury C. Biotechnology for improved product quality. World Leather 2002; 15(1):40-42.
45. McLaughlin GD, Highberger JH, O'Flaherty F et al. Some studies of the science and practice of bating. J Am Leather Chem Ass 1929; 24:339-79.
46. Wilson JA, Merrill HB. Activities of pancreatic enzyme used in bating upon different substrates, J Am Leather Chem Ass 1926; 21:50-53.
47. Quarck R. The bating problem. J Am Leather Chem Ass 1932; 27:532-41.
48. Stubbings R. Practical bating. The effect of bating variables on side and calf leather qualities. J Am Leather Chem Ass 1957; 52:298-311.
49. Gustavson KH. A new approach to the evaluation of action of proteinases on collagen in bating. J Am Leather Chem Ass 1949; 44:392-99.
50. Moore HN. Bating studies II. The comparative bating efficiency of bates of different biological sources. J Am Leather Chem Ass 1952; 47:110-27.
51. Uehara K, Toyoda H, Chonan Y et al. Observation of skins, hides and leather with scanning electron microscope, IV. Changes in pig skins during bating process. J Am Leather Chem Ass 1987; 82:33-40.
52. Marsal A, Cot J, De Castellar MD et al. On the recovery of natural fat and non ionic surfactant from sheepskin degreasing. J Am Leather Chem Ass 1998; 93:207-14.
53. Papp P, Toth K, Jancso J. Hung Teljes 3,325, 1972.
54. Jareckas G. Proizvod Primen Mikrobn Fermentn Prep 1975; 2:179.
55. Jareckas G, Shestakova IS. Proizvod Primen Mikrobn Fermentn Prep 1976; 3:277.
56. Yeshodha K, Dhar SC, Santappa M. Studies on the degreasing of skins using a microbial lipase. Leather Sci 1978; 25:77-86.
57. Muthukumaran N, Dhar SC. Comparative studies on the degreasing of skins using acid lipase and solvent with reference to the quality of finished leathers. Leather Sci 1982; 29:417-24.
58. Muthukumaran N, Dhar SC. Purification and characterization of Rhizopus nodosus acid lipase. Leather Sci 1983; 30:1-13.
59. Muthukumaran N, Dhar SC. Substrate specificity of Rhizopus nodosus acid lipase, Leather Sci 1983; 30:238-42.
60. Zhang X, Zhang Y. Test use of alkaline lipase in degreasing of pigskin. Pige Keji 1982; (7):40.
61. Pfleiderer E. Optimale Naßentfettung von Rohhaut, Blößen und wetblues. Leder 1983; 34:181-86.
62. Pfleiderer E, Taeger T, Wick G. Verfahren zur Naßentfettung von Hautmaterial. German Patent DE3,312,840, 1984.
63. Taegar T, Pfleiderer E. The analysis of enzymatic leather auxiliaries, J Soc Leather Technol Chem 1984; 68:105-08.
64. Galante YM, Formantici C. Enzyme applications in detergency and in manufacturing industries. Curr Org Chem 2003; 7:1399-422.
65. van Bailen JB, Li Z. Enzyme technology: an overview. Curr Opinion Biotechnol 2002; 13:338-44.
66. Kirk O, Borchert TV, Fuglsang CC. Industrial enzyme applications. Curr Opinion Biotechnol 2002; 13:345-51.
67. Auer N, Covington AD, Evans CS et al. Enzymatic removal of dung from hides. J Soc Leather Technol Chem 1999; 83:215-19.

68. Huang Q, Peng Y, Li X et al. Purification and characterization of an extracellular alkaline serine protease with dehairing function from Bacillus pumilus. Curr Microbiol 2003; 46:169-73.
69. Poza M. de Miguel T, Sieiro C et al. Characterization of a broad pH range protease of Candida caseinolytica. J Appl Microbiol 2001; 91:916-21.
70. Rao MB, Tanksale AM, Ghatge MS et al. Molecular and biotechnological aspects of microbial proteases. Microbiol Mol Biol Rev 1998; 62:597-635.
71. Martignone G, Monteverdi R, Roaldi M et al. Progress for leather makers. Leather 1997; 199(12):35-40.
72. Paul RG, Mohamed I, Davighi D et al. The use of neutral protease in enzymatic unhairing. J Am Leather Chem Ass 2001; 96:180-85.
73. Cantera CS. Hair saving unhairing process: Part 4. Remarks on the evolution of the investigations on enzyme unhairing. J Soc Leather Technol Chem 2001; 85:125-32.
74. Cantera CS, Goya L, Galarza B et al. Hair saving unhairing process: Part 5. Characterization of enzymatic preparations applied in soaking and unhairing processes. J Soc Leather Technol Chem 2003; 87:69-77.
75. Dayanandan A, Kanagaraj J, Sounderraj L et al. Application of an alkaline protease in leather processing: An ecofriendly approach. J Clean Prodn 2003; 11:533-36.
76. Thangam EB, Nagarajan T, Rajkumar G et al. Application of alkaline protease isolated from Alcaligenes faecalis for enzymatic unhairing in tanneries. J Am Leather Chem Ass 2001; 96:127-32.
77. Gehring AG, DiMaio GL, Marmer WN et al. Unhairing with proteolytic enzymes derived from Streptomyces Griseus. J Am Leather Chem Ass 2002; 97:406-11.
78. Pal S, Banerjee R, Chakraborty R et al. Application of a proteolytic enzyme in tanneries as a depilating agent. J Am Leather Chem Ass 1996; 91:59-63.
79. El Baba HAM, Covington AD, Davighi D. The effects of hair shaving on unhairing reactions: Part 1. The environmental impact. J Soc Leather Technol Chem 1999; 83:200-03.
80. El Baba HAM, Covington AD, Davighi D. The effects of hair shaving on unhairing reactions: Part 2. A new mechanism of unhairing. J Soc Leather Technol Chem 2000; 84:48-53.
81. Letourneau F, Soussotte V, Bressollier P et al. Keratinolytic activity of Streptomyces sp. S.K1-02: a new isolated strain. Lett Appl Microbiol 1998; 26(1):77-80.
82. Bressollier P, Letourneau F, Urdaci M et al. Purification and characterization of a keratinolytic serine proteinase from Streptomyces albidoflavus. Appl Environ Microbiol 1999; 65:2570-76.
83. Allpress JD, Mountain G, Gowland PC. Production, purification and characterization of an extracellular keratinase from Lysobacter NCIMB 9497. Lett Appl Microbiol 2002; 34(5):337-42.
84. Macedo AJ, Beys da Silva WO, Gava R et al. Novel keratinase from Bacillus subtilis S14 exhibiting remarkable dehairing capabilities. Appl Environ Microbiol 2005; 71:594-96.
85. Thanikaivelan P, Rao JR, Nair BU et al. Approach towards zero discharge tanning: Exploration of NaOH based opening up method. J Am Leather Chem Ass 2001; 96:222-33.
86. Schlosser L, Keller W, Hein A et al. The utilisation of a Lactobacillus culture in the beamhouse. J Soc Leather Technol Chem 1986; 70:163-68.
87. Hameed A, Keshavarz T, Evans CS. Effect of dissolved oxygen tension and pH on the production of extracellular protease from a new isolate of Bacillus subtilis K2, for use in leather processing. J Chem Tech Biotechnol 1999; 74:5-8.
88. Hameed A, Natt MA, Evans CS. Comparative studies of a new microbial bate and the commercial bate Oropon in leather treatment. J Indust Microbiol Biotechnol 1996; 17:77-79.
89. Ahmed MM, Gasmelseed GA. Application of an enzymatic bate from local materials. J Soc Leather Technol Chem 2003; 87:135-37.
90. Yongquan L. Sheep-pelt bating with acid protease. J Am Leather Chem Ass 2001; 96:398-400.
91. Mozersky SM, Allen OD, Marmer WN. Vigorous proteolysis: reliming in the presence of an alkaline protease and bating (post liming) with an extremophile protease. J Am Leather Chem Ass 2002; 97: 150-55.
92. Palop R, Marsal A, Cot J. Optimization of the aqueous degreasing process with enzymes and its influence on reducing the contaminant load. J Soc Leather Technol Chem 2000; 84:170-76.
93. Ivanova D, Costadinnova L, Todorova R. The influence of some parameters in leather degreasing on the lypolytic activity of lipase. J Soc Leather Technol Chem 2004; 88:161-63.
94. Addy VL, Covington AD, Langridge DA et al. Microscopy methods to study lipase degreasing: Part 2: A study of the interaction of ovine cutaneous adipocytes with lipase enzymes using microscopy. J Soc Leather Technol Chem 2001; 85:52-65.
95. Thanikaivelan P, Rao JR, Nair BU et al. Progress and recent trends in biotechnological methods for leather processing. Trends Biotechnol 2004; 22:181-88.
96. Thorstensen T. Pollution prevention and control for small tanneries. J Am Leather Chem Ass 1997; 92:245-55.
97. Crispim A, Mota M. Unhairing with enzymes. J Soc Leather Technol Chem 2003; 87:198-202.

98. Deselnicu M, Bratulescu V, Siegler M et al. A new enzyme process for improved yield and softer leather—technical note. J Am Leather Chem Ass 1994; 89:352-56.
99. Collighan RJ, Li X, Parry J et al. Transglutaminases as tanning agents for the leather industry. J Am Leather Chem Ass 2004; 99:293-98.
100. Rasmussen L. Wet blue enzymes—new treatment for area gain. World Leather 2002; 15(1):44-45.
101. Mitchell J, Ouellette D. Enzymes in retanning for cleaner blue stock. J Am Leather Chem Ass 1998; 93:255-59.
102. Thanikaivelan P, Rao JR, Nair BU. Development of a leather processing method in narrow pH profile. Part 1. Standardisation of unhairing process. J Soc Leather Technol Chem 2000; 84:276-84.
103. Saravanabhavan S, Aravindhan R, Thanikaivelan P et al. An integrated eco-friendly tanning method for the manufacture of upper leathers from goatskins. J Soc Leather Technol Chem 2003; 87:149-58.
104. Aravindhan R, Saravanabhavan S, Thanikaivelan P et al. A bio-driven lime and pickle free tanning paves way for greener garment leather production. J Am Leather Chem Ass 2004; 99:53-66.
105. Saravanabhavan S, Thanikaivelan P, Rao JR et al. Silicate enhanced enzymatic dehairing: A new lime-sulfide-free process for cowhides, Environ Sci Technol 2005; 39:3776-83.
106. Thanikaivelan P, Rao JR, Nair BU et al. Zero discharge tanning: A shift from chemical to biocatalytic leather processing. Environ Sci Technol 2002; 36:4187-94.
107. Saravanabhavan S, Aravindhan R, Thanikaivelan P et al. A Source Reduction Approach: Integrated bio-based tanning methods and the role of enzymes in dehairing and fibre opening, Clean Techn Environ Policy 2005; 7:3-14.
108. Thanikaivelan P, Bharath CK, Saravanabhavan S et al. Single step hair removal and fibre opening process: Simultaneous and successive addition of protease and α-amylase. J Am Leather Chem Ass (in press).
109. Thanikaivelan P, Rao JR, Nair BU et al. Biointervention makes leather processing greener: An integrated cleansing and tanning system. Environ Sci Technol 2003; 37:2609-17.
110. Saravanabhavan S, Aravindhan R, Thanikaivelan P et al. Green solution for tannery pollution: Effect of enzyme based lime-free unhairing and fibre opening in combination with pickle-free chrome tanning. Green Chem 2003; 5:707-14.
111. Saravanabhavan S, Thanikaivelan P, Rao JR et al. Natural leathers from natural materials: Progressing toward a new arena in leather processing, Environ Sci Technol 2004; 38:871-79.
112. Thanikaivelan P, Rao JR, Nair BU et al. Eco-friendly bioprocess for leather processing. US6708531, 2004.
113. Saravanabhavan S, Thanikaivelan P, Rao JR et al. An enzymatic beamhouse process coupled with semi-metal tanning and eco-benign post tanning leads to cleaner leather production, J Am Leather Chem Ass 2005; 100:174-86.
114. Saravanabhavan S, Thanikaivelan P, Chandrasekaran B et al. A new leather-making process for meeting eco-label standards: Processing of goatskins, J Am Leather Chem Ass 2006; 101:192-205.

CHAPTER 6

Degradation of Keratin- and Collagen-Containing Wastes by Enzyme Mixtures Produced by Newly Isolated Thermophylic Actinomycetes

Diana Braikova, Evgenia Vasileva-Tonkova,* Adriana Gushterova and Peter Nedkov

Abstract

Thermophilic actinomycetes isolated from Bulgarian and Antarctic soils were investigated for production of collagenolytic and keratinolytic enzymes during growth on keratin- and collagen-containing wastes used as a sole carbon source. Strains *Microbispora* sp. A-35 and *Thermoactinomyces* sp. E-15 that were isolated from Bulgarian soils produced highly stable "collagenase-like" enzymes with molecular weights between 46 and 140 kDa. Preliminary experiments showed that the enzyme mixture obtained by both strains is especially suitable for application in the pelt industry for "softening" the leathers. A simple and low-cost method was proposed for rapid and effective biodegradation of keratin- and collagen-containing wastes by four strains of the genus *Thermoactinomyces*, isolated from Bulgarian soils. The obtained microbial hydrolysates contained predominantly low-molecular peptides and amino acids, including essential ones, while the alkaline hydrolysate contained mainly peptides of higher molecular weight. Therefore, the obtained microbial products could be applied not only as fertilizers but even as protein additives to livestock and fish food. It was shown for first time that Antarctic actinomycetes were able to grow on keratin wastes producing keratinolytic enzymes. Of these, *Streptomyces flavis* 2BG (mesophilic) and *Microbispora aerata* 11A (thermophilic) possesed relatively high keratinase activity, which was stimulated by adding starch to the growth medium. The results showed that both strains are very promising for effective processing of keratinous wastes.

Introduction

Leather and fur plants, poultry plants, slaughterhouses, etc., generate huge amounts of waste by-products such as wool, feathers, bristle, bones, horns, hoofs, bones, hides and skins, which are mainly keratin and collagen proteins. In the near past, these wastes were baked and milled and then used as protein supplement to animal feedstuffs (animal flour). It was established, however, that this flour is the carrier of the unconventional infectious agent, the prion protein, the causative agent of a group of diseases called transmissible spongiform encephalopathies (TSE) that include mad cow disease, scrapie, kuru and Creutzfeld–Jakob disease. The pathogenic and infectious

*Corresponding Author: Evgenia Vasileva-Tonkova—Institute of Microbiology, Bulgarian Academy of Sciences, Acad. G. Bonchev str., bl. 26, 1113 Sofia, Bulgaria.
Email: evaston@yahoo.com

Enzyme Mixtures and Complex Biosynthesis, edited by Sanjoy K. Bhattacharya.

form of prion protein, PrP^{Sc}, is able to aggregate and form amyloid fibrils that are very stable and resistant to most disinfecting processes and common proteases.[1] Because animal flour is the carrier, the feeding of domestic animals was strictly prohibited in the countries of EU. Now, only the incineration of animal wastes is considered reliable way to break the spread of prions. Since such incineration has large expenses for transport, fuel, equipment, labour, etc., now these wastes are predominantly thrown away over controlled dung-hills. The hills however, take huge areas that could be used for agriculture, are sources of infections, and produce stench that creates serious ecological and sanitary problems. Therefore, an improved, more effective, and hopefully profitable utilization of these wastes is desirable, because they are a source of valuable products suitable for many applications. For this purpose, destruction of the rigid structure of keratins and collagens is necessary. Owing to their insoluble nature, these proteins are resistant to degradation by common proteolytic enzymes like trypsin, pepsin and papain because of intensive cross-linkage by disulfide bridges and other types of linkages.[2] Degradation of these wastes is usually achieved by thermal hydrolysis in dilute acid or base, or by enzymic digestion by specific proteases (keratinases and collagenases).[3-11] The usual proteolytic enzymes could be used after chemical treatment of feather meal at elevated temperature and high pressure. The hydrothermal treatment, in addition to being expensive, resulted in the destruction of certain essential amino acids like methionine, lysine and tryptophan which yielded a product with poor digestibility and variable nutrient quality.[12,13]

The use of microorganisms for obtaining soluble proteins and amino acids from keratin and collagen wastes is an attractive alternative method because it is ecologically sound and offers cheap and mild reaction conditions for production of valuable products.[14] Microbial degradation of these proteins is performed by specific proteases (keratinases and collagenases), or by proteases active under highly denaturing conditions destabilizing the molecules of recalcitrant proteins.[8,10,11,15,16] Microorganisms frequently produce trypsin and chymotrypsin-like serine or thiol proteases and subtilisin-like proteases with strong activity of broad specificity.[17-21] Nonspecific proteases participate in degradation of oligopeptides obtained from keratin and collagen by keratinases and collagenases and promote utilization of other proteins that occur in the habitats of microorganisms. Microbial collagenases and keratinases are distinguished by a big diversity of properties and could be used in fundamental scientific research and in biotechnology when there is a need for an effective, specific and fast hydrolysis of native and denatured keratin and collagen fibres of different types. Potential application of these enzymes include different fields such as medicine and pharmaceuticals, cosmetics, detergent and leather industries, developing cost-effective by-products for feed and fertilizers, biotransformation, etc.[11,22-24] Microbial keratinolytic and collagenolytic enzymes may have important use in biotechnological processes especially for waste bioconversion through the development of nonpolluting processes.[8] There have been some reports on the use of microorganisms and keratinolytic enzymes produced by them to degrade feather keratin more quickly and completely.[25-28] Recently there has been an increased interest in the production of biodegradable films, coatings and glues from keratinous waste products for compostable packaging, agricultural films or edible film applications.[29] Keratin structure is chemically modified and hydrolyzed to produce stable dispersions for such applications. Alternatively controlled hydrolysis of keratin using keratinases could offer environment-friendly technology. Recently, the broadspectrum bacterial keratinase PWD1 (Versazyme) was demonstrated to degrade completely prions from brain tissue of bovine spongiform encephalopathy (BSE) and scrapie-infected animals in the presence of detergents and heat treatment.[30,31]

A variety of different mesophilic strains, mainly bacteria and fungi, produce collagenolytic and keratinolytic enzymes.[8,10,11,15,32-35] Industrial-scale application of enzymes produced by mesophilic strains has been hampered because of limited thermal stability of the enzymes. Recent developments showed that thermophiles are a good source of novel enzymes that are of great industrial interest.[36] The main advantages of processes at higher temperatures are reduced risk of microbial contamination, lower viscosity, improved transfer rates, and improved solubility of substrates. Higher thermostability of the enzymes would be a significant advantage for their isolation, purification and storage. Until now, a limited number of studies have been reported on keratinolytic and collagenolytic enzymes produced by thermophilic microorganisms.[37-40] The use of thermophilic microorganisms for utilization of keratin and collagen wastes provide a unique opportunity due to the higher thermostability of secreted proteolytic enzymes and favourable changes in most physical properties of recalcitrant proteins and similarity to their structure pathogenic "prion" proteins at high temperatures. Increase in yields and activity of the enzymes is required to make these catalysts suitable for industrial applications. In addition, the mechanism regarding the enzymatic hydrolysis of insoluble proteins is highly complex and remains unknown. Therefore, isolation of new enzymes produced by newly isolated microorganisms including thermophiles seems to have good prospects. In this sense, attention has been paid recently to the thermophilic actinomycetes as possible producers of new thermostable enzymes.[41-44]

Collagenolytic Enzymes Produced by Thermophylic Actinomycetes Newly Isolated from Bulgarian Soils

Collagen is an insoluble structural protein found predominantly in skin, ligaments, bones, teeth, etc., that accounts for approximately 30% of the total weight of animal proteins and is produced in large quantities as a by-product in livestock industries. Generally, the ordinary proteinases do not digest the native triple-helical collagen type I; digestion is achieved by the collagen-specific, Zn^{2+}-dependent metalloproteinases (collagenases).[10,15] The process of collagen degradation produces peptides, which have been shown to have several biological activities of industrial and medical interest,[45] which lead to a wide variety of applications, e.g., an immunotherapeutic agent, a moisturizer for cosmetics, a preservative, and seasoning and dietary materials.[46-49] Disintegrating collagen fibers with high molecular fragments formation, collagenases fulfils the most important functions in a human or an animal organism by participating in transformation of the connective tissue in the process of its growth and morphogenesis, in cases of arthritis, tumour metastasis, and others. Collagenases, digesting different types of collagen, are excreted from tissues of vertebrates, crabs, flies grubs, insects' excretes, and are produced by microorganisms.

Based on previous screening for collagenolytic activity of thermophilic actinomycetes isolated from Bulgarian soils, two of them, determined as *Microbispora* sp. A-35 and *Thermoactinomyces* sp. E-15, were selected for production of collagenolytic enzymes.[50] Strains were cultivated for 48 hours at 55°C in liquid medium optimized for collagenase biosynthesis.[51] The produced enzymes were characterized and assessed for possible practical use.[52] The proteolytic activity was determined in Committee of Thrombolytic Agents (CTA) units with casein as a substrate.[53] One CTA unit was defined as the amount of enzyme releasing from the substrate 1 μeq tyrosine per minute at 37°C. The peptidase activity was measured towards different low molecular synthetic substrates: N_α-protected *p*-nitroanilides (*p*NA) of amino acids (N_α-glutaryl-L-Phe-*p*NA, N_α-benzoyl-L-Arg-*p*NA) and peptides (N_α-acetyl-Leu-Leu-Arg-pNA). The collagenolytic activity was assayed for 18 hours at 37°C in 50 mm Tris-HCl buffer in presence of 5 mm $CaCl_2$, pH 7.4 using acid soluble collagen type I (Sigma®) as a substrate and glycine as a standard.[54] One unit (U) of collagenolytic activity was defined as the amount of enzyme, which liberates from 10 mg collagen peptides, whose colour products with ninhydrin correspond to those obtained from 1 μmol glycine. The specific collagenolytic activity was determined against a synthetic substrate, Pz-Pro-Leu-Gly-Pro-D-Arg (Sigma®) (Pz, 4-phenylazobenzyloxycarbonyl), known as Pz-peptide and resembling the collagen structure.[55,56] The extent of specific cleavage (at the Leu-Gly peptide bond) was monitored by means of a standard peptide Pz-Pro-Leu (Bachem®) at 320 nm in presence of 0.1 M $CaCl_2$ at 25°C for 15 minutes.

Enzyme Production and Stability

The extracellular proteolytic and collagenase activities (total and specific) of A-35 and E-15 strains are shown in Figure 1. In comparison with other microbial collagenases,[57] the strains could be considered good producers of collagenolytic enzymes. The usual proteolytic enzymes are not able to cleave the native collagen triple-helix but quickly degrade it after denaturation or moderate

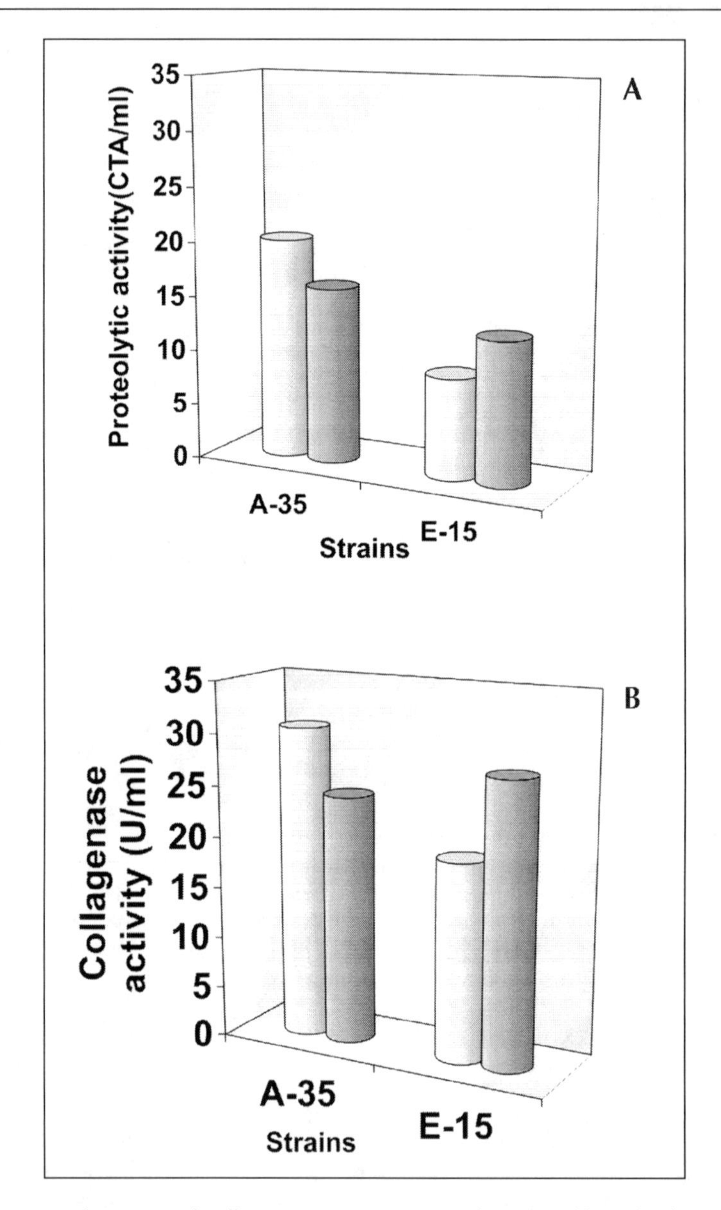

Figure 1. Proteolytic (A) and collagenase (B) activities in the cultural liquids of A-35 and E-15 strains expressed as total activity (white columns) and specific activity (grey columns).

hydrolysis of several peptide bonds in it.[58] Thus, the presence of proteolytic enzymes in a collagenase preparation is of critical importance for its successful application. Therefore, E-15 isolate seems to be more prospective for processing collagen-rich materials. The thermostability of produced enzymes was checked for a period of 30 minutes (Fig. 2). The proteolytic activity decreased with increased temperature and this process was sharper with A-35 than E-15, indicating that the latter strain is more suitable for industrial application. The enzymes produced by both strains were

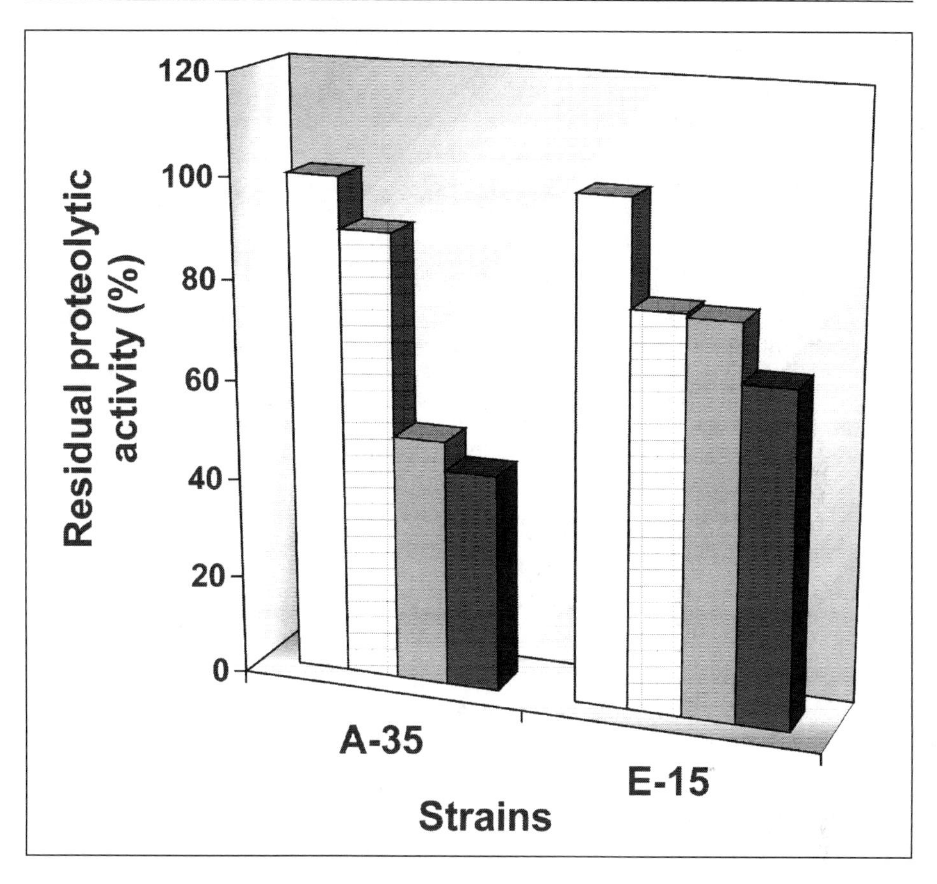

Figure 2. Residual proteolytic activity after treatment of enzyme preparations of A-35 and E-15 strains for 30 minutes at different temperatures. Control, white columns; 60°C, hatched columns; 65°C, grey columns; 70°C, black columns.

Table 1. Influence of multiple freezing/thawing on the extracellular enzyme activities of A-35 and E-15 Actinomycete strains

Cycles Freesing/Thawing	Proteolytic Activity (CTA/ml)		Collagenolytic Activity (U/ml)	
	Strain A-35	Strain E-15	Strain A-35	Strain E-15
0	3.21 ± 1.1	1.80 ± 1.5	16.69 ± 0.5	19.44 ± 0.6
1	3.63 ± 0.8	2.41 ± 0.2	15.29 ± 0.3	17.37 ± 0.3
2	4.03 ± 0.2	2.56 ± 0.1	12.08 ± 0.3	15.26 ± 0.2
3	3.90 ± 0.3	2.31 ± 0.2	11.78 ± 0.2	14.26 ± 0.3

relatively stable towards multiple cycles of freezing/thawing (Table 1). The precipitation of protein components with chilled ethanol (–25°C) is usually applied to remove carbohydrates, pigments and other by-products from the raw material. When the ethanol concentration ranged from 30 to

60%, 5-7% of the initial proteolytic activity was lost in the precipitate. At 70% ethanol, 70.6% of the proteolytic activity of E-15 and 86.4% of the proteolytic activity of A-35 were recovered from the precipitate. This indicates that proteins produced by both strains are strongly hydrated, i.e., they may have exposed on their molecular surface a large number of polar amino acid residues.

Substrate Specificity and Inhibitory Analysis

Molecular weights of enzymes produced by the strains were in the range from 46 to 140 kDa as was determined by chromatography on Sephadex G-100 column. Similar values have been obtained for enzymes produced by other actinomycete strains.[59,60] Substrate specificity of the enzymes was tested towards amino acid and peptide derivatives. It was established that the enzymes did not cleave N_α-benzoil-arginine-*p*-nitroanilide (N_α-benzoil-Arg-*p*NA), i.e., they are not typical serine proteases. Only the peptide substrate N_α-acetyl-Leu-Leu-Arg-pNA was cleaved to certain extent. The enzyme-substrate interaction is probably "multicenter"-one, which is typical for microbial proteases or thermolysin-like proteases.[61] The expressed specific collagenolytic activity, determined by PZ-peptide as a substrate, was 2.6-times (for E-15 strain), and 4-times (for A-35 strain) lower than that of clostridiopeptidase A.[62] So, the produced enzymes could be considered as "nonreal" collagenases, i.e., "collagenase-like" enzymes.

Different specific inhibitors were used in order to establish the type of produced enzymes. After treatment with PMSF a decrease with 10-16% in the collagenolytic activity of both strains was observed. The proteolytic activity of strain A-35 decreased with about 10% while that of strain E-15—with about 60%. It is evident that the collagenolytic component in both strains is not susceptible to the PMSF-treatment, whereas a part of the proteolytic enzymes of E-15 are susceptible to this inhibitor. The treatment with Ellman's reagent (DTNB, inhibitor of thiol proteases) leads to 25% loss of the proteolytic activity expressed by strain E-15. Chelate-forming agents (EDTA and *o*-phenantroline, inhibitors of metallo-proteinases, including collagenases) were applied in typical concentrations (5 mm for EDTA and 0.1 mm for *o*-phenantroline). The relatively high residual proteolytic activity, especially by strain E-15, confirms only a partial dependence of the proteolytic activity on the presence of metal ions. Taking into account the observed positive contribution of Ca^{2+} on the growth and the extracellular production, we may consider the produced enzymes as [Ca^{2+}]-dependent.

Possible Applications

Processing industries, and leather making in particular, cause adverse changes in the immediate environment and are, therefore, being challenged by society. Leather processing technology involves a series of operations, amongst which pretanning contributes to the major amount of pollution (approximately 70%). Sodium sulfide, lime and solid wastes generated as a result of pretanning are mainly responsible for increased biochemical oxygen demand (BOD), chemical oxygen demand (COD) and total dissolved solids (TDS).[63] The use of enzyme-based products is currently being explored for many areas of leather making in the development of nonpolluting processes.

Concentrated enzyme mixtures produced by the strains E-15 and A-35 were used in preliminary experiments for degradation of collagen-rich material (nontanned hide powder).[52] An increase with 30% of the ninhydrin-positive compounds was observed by strain E-15, and with 20% by strain A-35. Taking into account the mild action of the investigated enzyme preparations on collagen-containing substrates it turns out that they are especially suitable for usage in the pelt industry for softening the leathers. They could be applied also on stages preceding tanning and dyeing for controlling the extent of these procedures. We suppose that an extended application of the enzyme complexes might cause undesired changes in the properties and durability of the collagen-rich material.

Degradation of Keratin and Collagen Containing Wastes by Newly Isolated Thermoactinomycetes or by Alkaline Hydrolysis

Keratin forms a major component of the epidermis and its appendages viz. hair, feathers, nails, horns, hoofs, scales and wool. On the basis of secondary structural configuration, keratins have been classified into α (α-helix of hair and wool) and β (β-sheets of feathers).[64,65] The keratin fibrils in both configurations are twisted in a parallel manner to form micro and macro fibrils that ensure stability of the fiber.[66,67] Keratins are also grouped into hard and soft keratins according to the sulfur content. Hard keratins found in appendages like feathers, hair, hoofs and nails have a high content of disulfide bonds and are tough and inextensible. Whereas, soft keratins like skin and callus have low content of disulfide bonds and are more pliable.[7,64,68] Keratin-degrading enzymes, keratinases [EC 3.4.21/24/99.11], are mainly serine or metallo proteases. Since most of the purified keratinases known to date cannot completely solubilize native keratin,[41,69] their exact nature and uniqueness for keratinolysis is still an enigma in the world of proteases.

We isolated eleven thermophilic strains from soil enriched with hydrolyzed wool waste. Isolates were screened for growth and keratinase production in liquid medium supplemented with sheep skin and wool waste as a sole source of carbon and nitrogen.[70] The strains, designated as 4M, 3H, 8H and 4C, showed high levels of soluble proteins in the culture broth after 120 hours of growth and were selected for further work. They were identified as belonging to the genus *Thermoactinomyces*. Microbial hydrolysates were compared with those obtained by alkaline hydrolysis and an evaluation was made of their effectiveness and possible practical application.

Microbial and Alkaline Hydrolysis of Wool Waste

Table 2 shows the results of wool degradation by four selected strains after 120 hours of cultivation at 55°C using different concentrations of substrate. High levels of soluble protein in the growth

Table 2. Degradation of wool waste by Thermoactinomycetes

Strain	Starting Wool (%)	Dried Precipitate (g)	Total Protein in the Broth (mg/ml)	Solubilized Protein (%)
4M	0.5	0.11	4.90	99.0
	0.8	0.19	5.34	66.7
	1.5	1.22	3.19	21.2
	3.0	2.08	5.38	17.9
	10.0	3.01	5.98	5.9
3H	0.5	0.12	4.35	87.0
	0.8	0.26	5.70	71.2
	1.5	1.08	2.01	13.4
	3.0	2.09	4.23	14.1
	10.0	2.60	8.12	8.12
8H	0.5	0.12	4.90	98.0
	0.8	0.31	4.40	55.0
	1.5	1.19	2.76	18.4
	3.0	2.29	6.28	20.9
	10.0	2.76	5.73	5.7
4C	0.5	0.12	4.90	95.0
	0.8	0.31	5.25	65.7
	1.5	1.19	2.76	17.5
	3.0	2.29	5.50	19.3
	10.0	2.41	7.23	7.2

Table 3. Solubilization of sheep wool waste in 100 ml alkaline medium heated in autoclave at 120°C for 20 minutes

Sample No.	Wool Added (g)	Solubilized Wool (g)	Solubilized Wool (%)	Nonsoluble Residue (g)	Dry Matter in the Solution (%)	pH after Hydrolysis
1	0	0	0	0	1.8	13.20
2	2.5	2.5	100	0.38	4.3	12.90
3	5.0	4.7	94	0.89	6.5	11.65
4	7.5	6.7	89	1.54	8.5	10.78
5	10.0	8.2	82	3.01	10.0	10.56
6	12.5	9.4	75	4.50	11.2	10.38
7	15.0	9.7	65	7.15	11.5	10.26
8	17.5	9.9	57	10.71	11.7	10.23
9	20.0	8.8	44	15.21	10.6	10.07

medium were observed, which decreased with increasing the quantity of the added wool. The wool at high concentrations was finely dispersed and remained in solid state in the liquid phase giving a turbid solution or sol. These solid particles could represent hydrophobic peptides that are insoluble at pH 7 to 8, since adding acetic acid up to a 60% concentration turned the turbid supernatant into an almost clear solution. Search for new microbial keratinases is a difficult task because of the presence of different factors that could act synergistically on keratin digestion. Frequently, it happens that the wool becomes finely dispersed as a result of the microbial activity, but the extracellular keratinolytic activity is very low. One explanation of this paradox, digested wool by low enzyme activity, is that during the cultivation of microorganisms the factors acting on the keratin substrate could include more than only the keratinase activity. Namely, it was shown that some microorganisms produce compounds containing free thiol groups[40] which could split the disulfide bonds in the keratin molecule, thus allowing an easier degradation not only by the keratinase, but also by the ordinary proteolytic enzymes present. Another possibility is that keratinolytic enzymes or thiol groups could be located on the surface of microbial cells.

Alkaline hydrolysis of wool wastes was carried out in 0.15 M KOH—0.05 M NaOH in autoclave (Table 3) or by microwave heating (Table 4).[70,71] As can be seen, with respect to the percent solubilized wool and quantity used, best results were obtained with 5% wool mixture. Microwave heating is more economical and ecological one and minimizes thermal losses.[72] Microwave radiation can penetrate deep into the folding layers of keratin to destabilize the bonds

Table 4. Solubilization of sheep wool waste in 100 ml alkaline medium heated by microwaves at 100°C for 60 minutes

Sample No.	Wool Added (g)	Solubilized Wool (g)	Solubilized Wool (%)	Nonsoluble Residue (g)	Dry Matter in the Sol (%)	pH after Hydrolysis
1	0	0	0	0	1.8	13.2
2	2.5	2.5	100	0.18	4.3	13.16
3	5.0	4.9	98	0.88	6.7	12.56
4	7.5	6.4	85	1.10	8.2	12.02
5	10.0	8.1	81	1.9	9.9	11.25
6	12.5	9.2	74	4.3	11.0	11.08

rapidly thus promoting the forward hydrolysis of hog hair waste.[9] In our study, the waste material was solubilized by exposing it only for a short time (1 hour) to not very high alkalinity and it is believed that the possible undesired changes because of the alkaline medium may not occur. Even with these advantages, the majority of the obtained products after microwave hydrolysis of the keratin waste represented relatively long peptides. As was shown by chromatography on Sephadex G-25, microbial hydrolysates contained predominantly low-molecular peptides and amino acids indicating degradation of the major part of wool waste. Keratin degradation by the strains led to an increase in free amino acids such as asparagine, glycine, proline and lysine. Moreover, nutritionally essential amino acids such as phenylalanine and methionine, which are rare in wool keratin, were also produced in considerable amounts as microbial metabolites, especially by the strain 4M (Table 5). Similar observations were reported for native feather degradation by thermophylic *Fervidobacterium islandicum*.[40] Therefore, the obtained products could be suitable not only as fertilizers but even as protein additives to livestock and fish food.

Possible Applications

On the basis of the results obtained, a simple and low-cost method was proposed for rapid and effective biodegradation of keratin wastes by a mixed culture of the four new isolated thermoactinomycete strains. The low molecular compounds are continuously removed during the microbial cultivation by ultrafiltration, then concentrated and spray-dried. Moreover, if the broth medium is forced to move through the waste by peristaltic pump, no milling or other prior mechanical treatment of keratin wastes will be required, which will substantially reduce processing costs. We are convinced that the microbial method of degradation is more useful than microwave alkaline hydrolysis since the first could be used for preparation of not only fertilizers but also of animal feed additives.

The final waste product after alkaline hydrolysis in autoclave contained 75-80% water-soluble materials: peptides, amino acids, lipids, some carbohydrates, salts, dyes, potassium ions, and 20-25% partially degraded highly dispersed keratin. We consider the alkaline hydrolysis a method

Table 5. Amino acid content of the alkaline and microbial hydrolysates (residues per 1000)

Amino Acid	Native Wool	Microwave Hydrolysate	Autoclave Hydrolysate	Strain 4M	Strain 3H	Strain 8H	Strain 4C
Asp	62	75	95	81	204	106	93
Thr	60	36	80	41	13	21	10
Ser	93	64	30	93	35	29	19
Glu	158	167	139	142	66	139	100
Pro	38	40	68	44	218	47	45
Gly	90	171	230	126	262	534	507
Ala	44	77	86	68	102	27	42
Cys	108	-	-	-	-	-	-
Val	64	90	32	36	4	7	12
Met	4	2	9	20	3	4	11
Ile	33	32	4	18	2	3	5
Leu	110	68	13	36	4	7	15
Tyr	26	34	49	50	7	11	10
Phe	32	42	21	70	14	20	47
Lys	25	12	18	91	14	20	48
His	24	42	63	55	6	11	18
Arg	29	48	62	29	46	14	18

suitable for obtaining harmless and valuable fertilizer for agriculture since the preparation is easily obtained and improves soil characteristics and positively influences microbial soil populations.[71] If this utilization of the keratin wastes is used, the environment around the leather and fur plants will be influenced positively and the expenses of removing the wastes to controlled dung-hills and part of the expenses for sustaining the latter will be saved.

Degradation of Keratin-Containing Wastes by Actinomycete Strains Newly Isolated from Antarctic Soils

The microflora of penguins' excrement and glaciers in the region of the Livingston Island, Antarctica is characterized by the presence of different ecological types of microorganisms: thermophils, mesophils, facultative psychrophils and typical psychrophils.[73,74] The presence of thermophilic actinomycetes in Antarctica was unexpected for us, but is a fact. It could be a result of pollution of this continent.[75] There are very large deposits of coal in Antarctica, which indicates that rich vegetation existed until glaciation, probably beginning 50 million years ago.[76] On this basis, another hypothesis could be that the termophilic microorganisms have existed since the Paleozoic and Mesozoic eras when Antarctica was warm, and their spores have survived (in cryanabiotic state) until now. We isolated nine thermophilic and ten mesophilic actinomycete strains from Antarctic soil samples and screened them for keratinolytic activity during growth on wool waste as a sole source of carbon and nitrogen.[77] Keratinase activity was determined by the modified method of Cheng et al (1995).[78] The increase in absorbance at 280 nm was converted into keratinase units (1KU = 0.1 absorbance increase for 1 hours). Two of the strains, the mesophilic *Streptomyces flavis* 2BG and the thermophilic *Microbispora aerata* 11A, showed relatively high levels of keratinase activity and soluble protein in the culture broth and were selected for more detail investigations.

Enzyme Production and Reduction of Disulphide Bonds in Keratin Substrate

The strains were cultivated for 10 days at 55°C (*Microbispora aerata*) or 28°C (*Streptomyces flavis*) in mineral salts medium with wool waste as the sole source of carbon and nitrogen. Maximum keratinase activity was observed on day 5 of cultivation for the strain 11A and on day 8 for the strain 2BG (Fig. 3B). The increase of keratinase activity was associated with an increase of soluble protein in the culture broth (Fig. 3A). At this time, maximum concentration of SH-groups in the broth medium was observed (Fig. 3C). A simultaneous cleavage of disulphide bonds during microbial growth has been reported for *Streptomyces fradiae*,[79] and *Streptomyces pactum*.[80]

Microbial keratinases are predominantly extracellular when grown on keratinous substrates; however, a few cell-bound[8,38,40,81] and intracellular keratinases have also been reported.[8,82] The intracellular fraction in most of these reports mainly contributes to disulfide reductases, sulfite or thiosulfate that synergistically assists the extracellular keratinases to degrade keratin by reducing the disulfide bonds of keratin. There are two steps in keratinolysis: sulfitolysis or reduction in disulfide bonds and proteolysis. It may be speculated that sulfitolysis requires either the presence of live cells;[69,83] reductants like sodium sulfite, DTT, mercaptoethanol, glutathione, cysteine and thioglycolate;[8] or disulfide reductases,[20,69] which act in co-operation with keratinolytic proteases to bring about complete degradation of keratin. However, the order of these events and their exact nature are still debatable.

Keratinases are largely produced in a basal medium with a keratinous substrate.[41] In most cases, keratin serves as the inducer; however, soy meal is also known to induce enzyme production.[34,78] Most of the reports available on keratinases group them as inducible enzymes; however, a few constitutive keratinases have also been reported.[84,85] Keratinase activity was not detected during growth of both tested strains in media without keratin. This indicated that the major regulatory mechanism for keratinase synthesis by these strains is substrate induction.

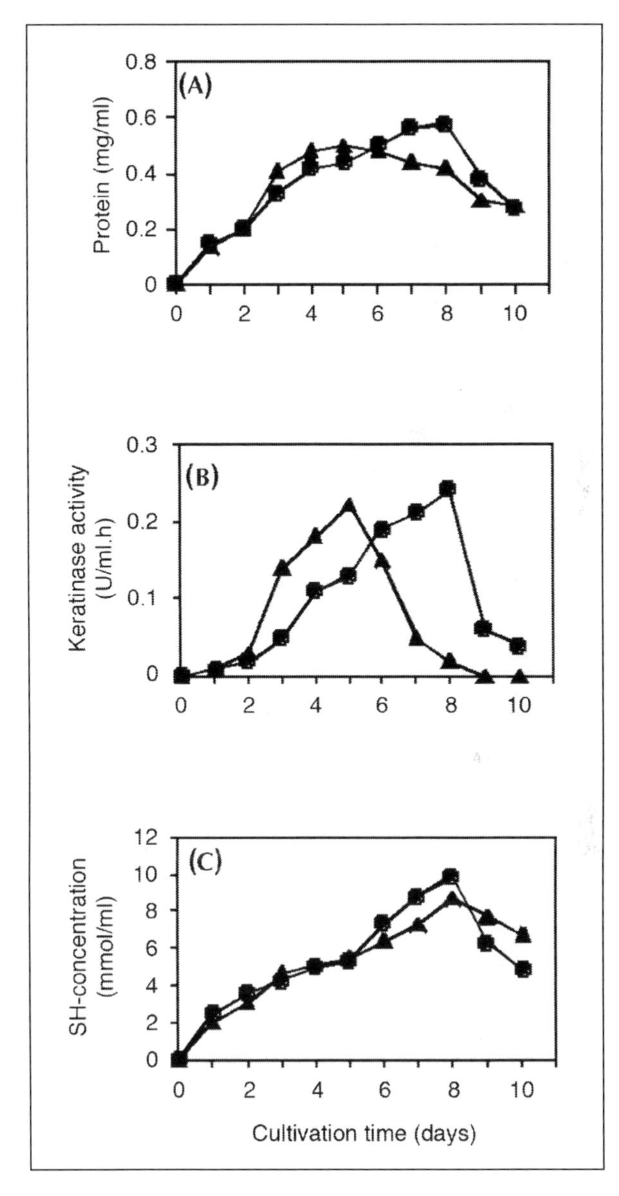

Figure 3. Soluble protein content (A), keratinase activity (B) and concentration of free thiol groups (C) for *Streptomyces flavis* 2BG A (●) and *Microbispora aerata* 11A (▲). The strains were grown in mineral medium (pH 7.2), supplemented with 6 g/L wool as the sole source of carbon and nitrogen. Mean values of three determinations are given.

Influence of Starch on Keratinase Production

Strains 2BG and 11A were tested for keratinase activity in the presence of increasing amounts of starch as the carbon source. Since starch is a main component in the media for isolation of actinomycetes,[86] it was added in order to stimulate the keratinase secretion. Moreover, it is a cheap and widely accessible product. After 5 days of cultivation, 6-fold higher activity was observed in presence

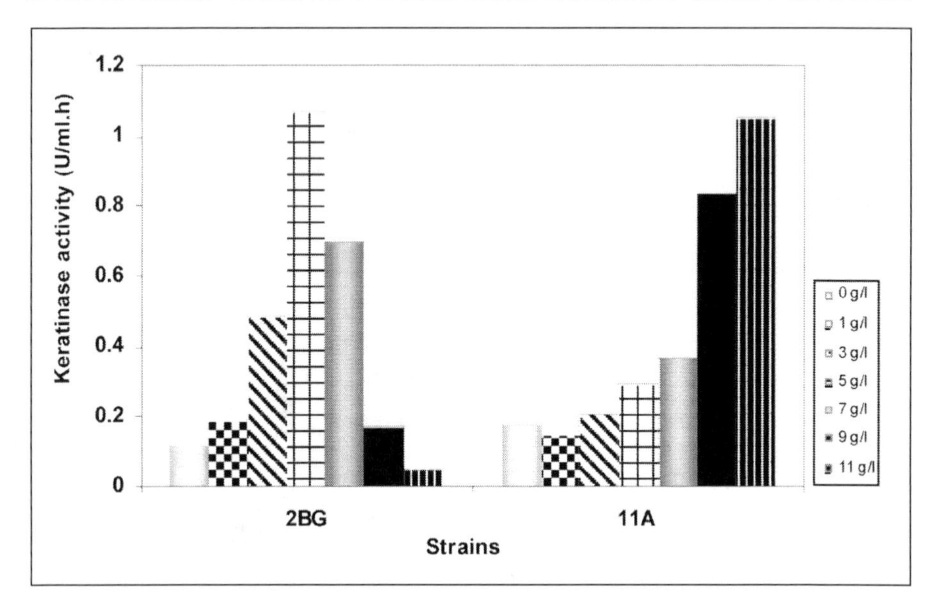

Figure 4. Effect of the addition of different concentrations of starch on keratinase production by *Streptomyces flavis* 2BG and *Microbispora aerata* 11A. Concentrations of starch are indicated in the legend. Mean values of three experiments are given.

of 11 g starch per liter for the strain 11A, and 9-fold higher for the strain 2BG in presence of 5 g starch per liter (Fig. 4). Both strains belong to different genera with different metabolism and have different dependence on the growth temperature. Because of that, they did not respond in the same way with increasing starch concentration. The concentration of SH-groups was found to decrease with increasing starch concentration. Most probably in this case, the disulphide bond reduction took place in the cell-bound redox system situated on the surface of the cell. This mechanism requires a close contact between the cell and the insoluble keratinous substrate. Similar observations have been reported for disulphide bond reduction by *Streptomyces pactum*.[83] So far no reports in the literature were found about the ability of Antarctic actinomycetes to grow on keratin wastes and to produce keratinolytic enzymes. The increase of keratinase activity in presence of starch in the growth medium of selected Antarctic actinomycete strains indicated considerable biotechnological potential for use in the leather industry and in the processing of keratinous wastes.

Acknowledgements

These studies were financially supported by Project SS-1210 of the Bulgarian Fund for Science and Technology.

References

1. Tsiroulnikov K, Rezai H, Bonch-Osmolovskaya E et al. Hydrolysis of the Amyloid Prion Protein and Nonpathogenic Meat and Bone Meal by Anaerobic Thermophilic Prokaryotes and Streptomyces Subspecies. J Agric Food Chem 2004; 52:6353-60.
2. Arai K, Naito S, Dang VB et al. Cross-Linking structure of keratin. VI. Number, type, and location of disulfide cross linkages in low-sulfur protein of wool fiber and their relation to permanent set. J Appl Polym Sci 1996; 60:169-179.
3. Kida K, Morimura S, Noda J et al. Enzymatic hydrolysis of the horn and hoof of cow and buffalo. J Ferment Bioengin 1995; 80:478-484.
4. Shih JCH. 1993 Recent developments in poultry waste digestion and feather utilization—a review. Poiltry Science 72, 1617-1620.

5. Dalev P, Vassileva E, Mark JE, Fakirov S. Enzymatic degradation of formaldehyde-crossing gelatin. Biotechnol Tech 1998; 12:889-892.
6. Morimura S, Nagata H, Uemura Y et al. Development of an effective process for utilization of collagen from livestock and fish waste. Process Biochem 2002; 37:1403-1412.
7. Barone JR, Schmidt WF. Effect of formic acid exposure on keratin fiber derived from poultry feather biomass. Biores Technol 2006; 97:233-242.
8. Onifade AA, Al-Sane NA, Al-Mussallam AA, Al-Zarbam S. Potential for biotechnological application of keratin-degrading microorganisms and their enzymes for nutritional improvement of feathers and other keratins as livestock feed resources. Biores Technol 1998; 66:1-11.
9. Jou CJG, Chen YS, Wang HP et al. Hydrolytic dissociation of hog-hair by microwave radiation. Bioresour Technol 1999; 70:111-113.
10. Harrington DJ. Bacterial collagenases and collagen-degrading enzymes and their potential role in human disease. Infect Immun 1996; 64:1885-91.
11. Gupta R, Ramnani P. Microbial keratinases and their prospective applications: an overview. Appl Microbiol Biotechnol 2006; 4:1-13. DOI: 10.1007/s00253-005-0239-8.
12. Papadopoulos MC. Effect of processing on high-protein feedstuffs: a review. Biol Wastes 1989; 29:123-138.
13. Wang X, Parsons CM. Effect of processing systems on protein quality of feather meals and hog hair meals. Poultry Sci 1997; 76:491-496.
14. Grazziotin A, Pimentel FA, de Jong EV, Brandelli A. Nutritional improvement of feather protein by treatment with microbial keratinase. Animal Feed Sci Technol 2006;126:135-144.
15. Demina NS, Lysenko SV. Collagenolytic enzymes synthesized by microorganisms. Review. Mikrobiologiia (Russian) 1996; 65:293-304.
16. Han XQ, Damodaran S. Stability of protease Q against autolysis and in sodium dodecyl sulphate and urea solutions. Biochem Biophys Res Commun 1997; 240:839-843.
17. Ferrero MA, Castro GR, Abate CM et al. Thermostable alkaline proteases of Bacillus licheniformis MIR 29: isolation, production and characterization. Appl Microbiol Biotechnol 1996; 45:327-332.
18. Evans KL, Crowder J, Miller ES. Subtilisins of Bacillus spp. hydrolyze keratin and allow growth on feathers. Can J Microbiol 2000; 46:1004-1011.
19. Rozs M, Manczinger L, Vagvolgyi Cs, Kevei F. Secretion of a trypsin-like thiol protease by a new keratinolytic strain of Bacillus licheniformis. FEMS Microbiol Lett 2001; 205:221-224.
20. Yamamura S, Morita Y, Hasan Q et al. Keratin degradation: a cooperative action of two enzymes from Stenotrophomonas sp. Biochem Biophys Res Commun 2002; 294:1138-1143.
21. Najafi MF, Deobagkar D, Deobagkar D. Potential application of protease isolated from Pseudomonas aeruginosa PD100. Electronic J Biotechnol 2005; 8:197-203.
22. Gupta R, Beg QK, Lorenz P. Bacterial alkaline proteases: Molecular approaches and industrial applications. Appl Microbiol Biotechnol 2002; 59:15-32.
23. Friedrich J, Kern S. Hydrolysis of native proteins by keratinolytic protease of Doratomyces microsporus. J Mol Catal B: Enzymatic 2003; 21:35-37.
24. Macedo AJ, da Silva WO, Gava R et al. Novel keratinase from Bacillus subtilis S14 exhibiting remarkable dehairing capabilities. Appl Environ Microbiol 2005; 71:594-596.
25. Sandali S, Brandelli A. Feather keratin hydrolysis by a Vibrio sp. strain kr2. J Appl Microbiol 2000; 89:735-743.
26. Ichida JM, Krizova L, LeFevre CA et al. Bacterial inoculum enhances keratin degradation and biofilm formation in poultry compost. J Microbiol Methods 2001; 47:199-208.
27. Kim JM, Lim WJ, Suh HJ. Feather-degrading Bacillus species from poultry waste. Process Biochem 2001; 37:287-291.
28. Singh CJ. Optimization of an extracellular protease of Chrysosporium keratinophilum and its potential in bioremediation of keratinic wastes. Mycopathologia 2003; 156:151-156.
29. Schrooyen PMM, Dijkstra PJ, Oberthur RC et al. Partially carboxymethylated feather keratins. 2. Thermal and mechanical properties of films. J Agric Food Chem 2001; 49:221-230.
30. Langeveld JP, Wang JJ, Van de Wiel DF et al. Enzymatic degradation of prion protein in brain stem from infected cattle and sheep. J Infect Dis 2003; 188:1782-1789.
31. Chen C-Y, Rojanatavorn K, Clark AC, Shih JCH. Characterization and enzymatic degradation of Sup35NM, a yeast prion-like protein. Protein Sci 2005; 14:2228-2235.
32. Thys RCS, Lucas FS, Riffel A et al. Characterization of a protease of a feather-degrading Microbacterium species. Lett Appl Microbiol 2004; 39:181.
33. Friedrich J, Gradišar H, Vrecl M, Pogačnik A. In vitro degradation of porcine skin epidermis by a fungal keratinase of Doratomyces microsporus. Enzyme Microbial Technol 2005, 36:455-460.

34. Gradisar H, Friedrich J, Krizaj I, Jerala R. Similarities and specificities of fungal keratinolytic proteases: comparison of keratinases of Paecilomyces marquandii and Doratomyces microsporus to some known proteases. Appl Environ Microbiol 2005, 71:3420-3426.
35. Werlang PO, Brandelli A. Characterization of a Novel Feather-Degrading Bacillus sp. Strain. Appl Biochem Biotechnol 2005; 120:71-80.
36. Bruins ME, Janssen AEM, Boom RM. Thermozymes and their applications: a review of recent literature and patents. Appl Biochem Biotechnol 2001; 90:155-186.
37. Atalo K, Gashe BA. Protease production by a thermophilic Bacillus species (P-001A), which degrades various kinds of fibrous proteins. Biotechnol Lett 1993; 15:1151-56.
38. Riesen S, Antranikian G. Isolation of Thermoanaerobacter keratinophilus subsp. nov., a novel, thermophilic, anaerobic bacterium with keratinolytic activity. Extremophiles 2001; 5:399-408.
39. Tsuruoka N, Nakayama T, Ashida M et al. Collagenolytic Serine-Carboxyl Proteinase from Alicyclobacillus sendaiensis Strain NTAP-1: Purification, Characterization, Gene Cloning, and Heterologous Expression. Applied Environ Microbiol 2003; 69:162-169.
40. Nam G-W, Lee D-W, Lee H-S et al. Native-feather degradation by Fervidobacterium islandicum AW-1, a newly isolated keratinase-producing thermophilic anaerobe. Arch Microbiol 2002; 178:538-547.
41. Ignatova Z, Gousterova A, Spassov G, Nedkov P. Isolation and partial characterization of extracellular keratinase from a wool degrading thermophilic actinomycete strain Thermoactinomyces candidus. Can J Microbiol 1999; 45:217-222.
42. Chitte RR, Nalawade VK, Dey S, Sey S. Keratinolytic activity from the broth of a feather-degrading thermophilic Streptomyces thermoviolaceus strain SD8. Lett Appl Microbiol 1999; 28:131-136.
43. Mohamedin AH. Isolation, identification and some cultural conditions of a protease-producing thermophilic Streptomyces strain grown on chicken feather as a substrate. Internat Biodet Biodegr 1999; 43:13-21.
44. Szabo L, Benedek A, Szabo ML, Barabas G. Feather degradation with a thermotolerant Streptomyces graminofaciens strain. World J Microbiol Biotechnol 2000; 16:252-255.
45. Ravanti L, Kähäri VM. Matrix metalloproteinases in wound repair. Int J Mol Med 2000; 6:391-407.
46. Ku G, Kronenberg M, Peacock DJ et al. Prevention of experimental autoimmune arthritis with a peptide fragment of type II collagen. Eur J Immunol 1993; 23:591-599.
47. Khare SD, Krco CJ, Griffiths MM et al. Oral administration of an immunodominant human collagen peptide modulates collagen-induced arthritis. J Immunol 1995; 155:3653-59.
48. Gaffney PJ, Edgell TA, Dawson PA et al. A pig collagen peptide fraction. A unique material for maintaining biological activity during lyophilization and during storage in the liquid state. J Pharm Pharmacol 1996; 48:896-898.
49. Honda S. Dietary use of collagen and collagen peptides for cosmetics. Food Style 1998; 21:54-60.
50. Gousterova A, Lilova A, Trenev M et al. Thermophylic actinomycetes as producers of collagenase. Compt Rend Acad Bulg Sci 1998; 51:71-74.
51. Christov P, Goushterova A, Goshev I et al. Optimization of the biosynthetic conditions for collagenase production by certain thermophilic actinomycete strains. Compt Rend Acad Bulg Sci 2000; 53:115-118.
52. Goshev I, Gousterova A, Vasileva-Tonkova E, Nedkov P. Characterization of the enzyme complexes produced by two newly isolated thermophylic actinomycete strains during growth on collagen-rich materials. Process Biochem 2005; 40:1627-31.
53. Nedkov P. A preparative method for producing partially purified alkaline protease. Commun Dept Chem Bulg Acad Sci 1986; 19:246-253.
54. Endo A. Novel collagenase "discolysin" and production method thereof. US Patent No 4624924.
55. Wünsch E, Heindrich HG. Zur quantitativen Bestimmung der Kollagenase. Hoppe-Seyler's Z Physiol Chem 1963; 333:149-151.
56. Chikuma T, Ishii Y, Kato T. Highly sensitive assay for PZ-peptidase activity by high-performance liquid chromatography. J Chromatogr 1985; 348:205-212.
57. Lund T, Granum PE. The 105-kDa protein component of Bacillus cereus nonhaemolytic enterotoxin (Nhe) is a metalloprotease with gelatinolytic and collagenolytic activity. FEMS Microbiol Lett 1999; 178:355-361.
58. Stryer L. Connective-tissue proteins: collagen, elastin and proteoglycans. In: Stryer L, ed. Biochemistry. 2nd Ed. San Francisco: W.H. Freeman and Company, 1981:185-198.
59. Rippon JW, Lorintz AL. Collagenase activity of Streptomyces (Nocardia) madurae. J Invest Dermatol 1964; 43:483-486.
60. Endo A, Murakawa S, Shimizu H, Shiraishi Y. Purification and properties of collagenase from a Streptomyces species. J Biochem (Tokyo)1987; 102:163-170.
61. Rao MB, Tanksale AM, Ghatge MS, Deshpande VV. Molecular and biotechnological aspects of microbial proteases. Microbiol Mol Biol Rev 1998; 62:597-635.

62. Van Wart HE. Clostridium collagenases. In: Barrett AJ, Rawlings ND, Woessner JF, eds. Handbook of Proteolytic Enzymes. London: Academic Press, 1998:1098-1102.
63. Thanikaivelan P, Rao JR, Nair U, Ramasami T. Progress and recent trends in biotechnological methods for leather processing. Trends Biotechnol 2004; 22:181-188.
64. Voet D, Voet JG. Three-dimensional structure of proteins. In: Stiefel J, ed. Biochemistry, 2nd ed. New York: Wiley, 1995:154-156.
65. Akhtar W, Edwards HGM. Fourier-transform Raman spectroscopy of mammalian and avian keratotic biopolymers. Spectrochim Acta 1997; 53:81-90.
66. Kreplak L, Doucet J, Dumas P, Briki F. New aspects of the α-helix to β-sheet transition in stretched hard α-keratin fibers. Biophys J 2004; 87:640-647.
67. Zerdani I, Faid M, Malki A. Feather wastes digestion by new isolated strains Bacillus sp. in Morocco. Afr J Biotechnol 2004; 3:67-70.
68. Schrooyen PMM, Dijkstra PJ, Oberthur RC et al. Partially carboxymethylated feather keratins. 2. Thermal and mechanical properties of films. J Agric Food Chem 2001; 49:221-230.
69. Ramnani P, Singh R, Gupta R. Keratinolytic potential of Bacillus licheniformis RG1: structural and biochemical mechanism of feather degradation. Can J Microbiol 2005; 51:191-196.
70. Gousterova A, Braikova D, Goshev I et al. Degradation of keratin and collagen containing wastes by newly isolated thermoactinomycetes or by alkaline hydrolysis. Lett Applied Microbiol 2005; 40:335-340.
71. Nustorova M, Braikova D, Gousterova A et al. Chemical, microbiological and plant analysis of soil fertilized with alkaline hydrolysate of sheep's wool waste. World J Microbiol Biotechnol 2006; 22:383-390.
72. Jones DA, Lelyveld TP, Mavrofidis SD et al. Microwave heating applications in environmental engineering-a review. Resour Conserv Recycling 2002; 34:75-90.
73. Gushterova A, Noustorova M, Tzvetkova R et al. Investigations of the microflora in penguin's excrements in the Antarctic. Bulgarian Antarctic Research, Life Sciences. Vol 2. Sofia-Moscow: Pensoft Pub, 1999:1-7.
74. Noustorova M, Gushterova A, Tzvetkova R, Chipeva V. Investigations of the microflora in glaciers from the Antarctic. Bulgarian Antarctic Research, Life Sciences. Vol 2. Sofia-Moscow: Pensoft Pub, 1999:8-12.
75. Agre NS. 1986 Taxonomy of thermophilic actinomycetes. PhD Thesis. Moscow: Puschino, 1986:5-10 (in Russian).
76. Encyclopedia Britannica 2004 Antarctica. In: Encyclopedia Britannica, from Deluxe Edition 2004 CD-ROM.
77. Gushterova A, Vasileva-Tonkova E, Dimova E et al. Keratinase production by newly isolated Antarctic actinomycete strains. World J Microbiol Biotechnol 2005; 21:831-834.
78. Cheng SW, Hu HM, Shen SW et al. Production and characterization of keratinase of a feather-degrading Bacillus licheniformis PWD-1. Biosci Biotechnol Biochem 1995; 59:2239-2243.
79. Kunert J, Stransky Z. Thiosulfate production from cystine by keratinolytic prokaryote Streptomyces fradiae. Arch Microbiol 1988; 150:600-601.
80. Bockle B, Galunsky B, Muller R. Characterization of a keratinolytic serine proteinase from Streptomyces pactum DSM 40530. Appl Environ Microbiol 1995; 61:3705-3710.
81. Friedrich AB, Antranikian G. Keratin degradation by Fervidobacterium pennavorans, a novel thermophilic anaerobic species of the order thermotogales. Appl Environ Microbiol 1996; 62:2875-2882.
82. El-Naghy MA, El-Ktatny MS, Fadl-Allah EM, Nazeer WW. Degradation of chicken feathers by Chrysosporium georgiae. Mycopathologia 1998; 143:77-84.
83. Bockle B, Muller R. Reduction of disulphide bonds by Streptomyces pactum during growth on chicken feathers. Appl Environ Microbiol 1997; 63:790-792.
84. Gassesse A, Hatti-Kaul R, Gashe BA, Mattiasson B. Novel alkaline proteases from alkaliphilic bacteria grown on chicken feather. Enzyme Microbial Technol 2003; 32:519-524.
85. Manczinger L, Rozs M, Vagvolgyi Cs, Kevei F. Isolation and characterization of a new keratinolytic Bacillus licheniformis strain. World J Microbiol Biotechnol 2003; 19:35-39
86. Kosmachev AE. Thermophilic actinomycetes and their antagonistic properties. Ph.D Thesis. Moscow: Inst Mikrobiol 1954 (in Russian).

CHAPTER 7

Production of Enantiomerically Pure Pharmaceutical Compounds Using Biocatalysts

Ülkü Mehmetoğlu,* Emine Bayraktar, Çiğdem Babaarslan

Abstract

The production of single enantiomers of chiral intermediates has become increasingly important, especially in pharmaceutical industry. The number of the drugs in single-enantiomer form is growing annually. While one of the enantiomers has a desired activity, the other has a toxic effect or no activity. During the last decade, the importance of enzyme-catalyzed reactions has been well recognized as a promising method for the preparation of enantiomerically pure compound. Asymmetric synthesis and the resolution of the racemate are the methods used to prepare enantiomers using enzyme. Asymetric synthesis is the conversion of prochiral substrate into a chiral product. There are two methods of resolution of the racemate enzymetically: kinetic resolution and dynamic kinetic resolution. The success of the kinetic resolution is dependent on the different reaction rates of the two enantiomers. In general the maximum yield of kinetic resolution is only 50%, which is economically unattractive, but this problem can be overcome for instance by achieving dynamic kinetic resolution in which the unreacted enantiomer is continuously racemised.

Isolated enzymes and whole-cell enzymes which are used in chemo-enzymatic synthesis are efficient catalysts under mild conditions. The reactions achieved by biocatalysts are faster and harmless to environment. Whole-cell enzymes used as biocatalyst have some advantages to isolated enzymes. While enzyme isolation and purification is very expensive and hard, whole-cell enzymes are inexpensive and available in the desired quantities. Especially, whole-cell enzymes are usually used in cofactor-dependent enzymatic reactions in order to allow a cheap regeneration of the cofactor. In contrast to pure enzymes, whole-cells systems (bacteria, fungi) can be available in sufficient quantities. In this chapter there are some production examples using isolated enzymes and whole-cell enzymes. In addition, the effects of the important parameters on the enantioselectivity are discussed.

Introduction

Because of biological advantages in recent years, production of single enantiomer is important. There is an increased demand for optically pure enantiomers as building blocks for pharmaceuticals. The number of the drugs in single-enantiomer form is growing annually. While one of the enantiomers has a desired activity, the other has a toxic effect or no activity. For example, (S) fluotexine

*Corresponding Author: Ülkü Mehmetoğlu—Ankara University, Faculty of Engineering, Department of Chemical Engineering, 06100 Tandogan, Ankara, Turkey; and Ankara University, Biotechnology Institiute, Ord. Prof. Dr. Sevket Kansu Binası, 06500 Besevler, Ankara, Turkey. Email: mehmet@eng.ankara.edu.tr

Enzyme Mixtures and Complex Biosynthesis, edited by Sanjoy K. Bhattacharya.

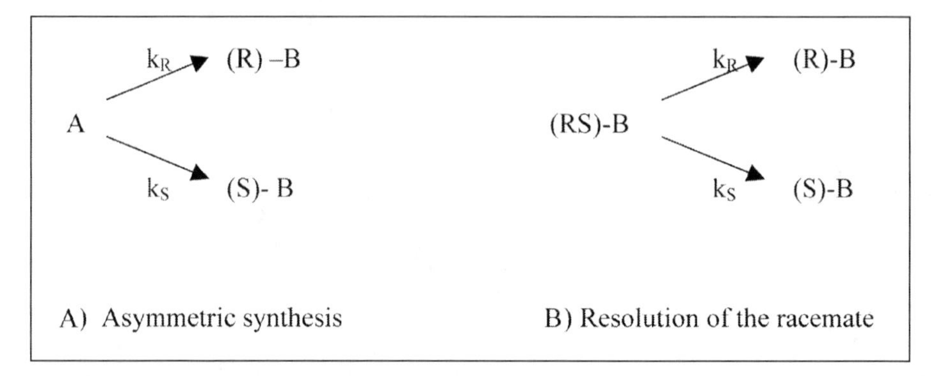

Figure 1. The preparation of enantiomers.

shows remarkable therapeutic effects in preventing migraines while the racemic drug has no effect. Biotechnologies and biocatalysts are rapidly expanding fields to produce and to purify chiral intermediates.[1] During the last decade, the importance of enzyme-catalyzed reactions has been well recognized as a promising method for the preparation of enantiomerically pure compound.[2]

Two methods are used for the preparation of enantiomers using enzyme (Fig. 1):[3]

a. Asymmetric synthesis.
b. The resolution of the racemate.

Optically pure materials can be synthesied by of transformating the prochiral substrate into a chiral product with asymmetric synthesis. Asymmetric synthesis generally involves redox processes (microbial oxidations and microbial reductions) or condensation reactions.[4]

There are two methods of resolution of the racemate enzymetically: kinetic resolution and dynamic kinetic resolution. Due to the chirality of the active site of the enzyme, one enantiomer fits better into the active site than its counterpart and is converted at a higher rate, resulting in a kinetic resolution of the racemate.[5] The success of this method depends on reacting two enantiomers at different rates with a chiral entity which may be an enzyme or a microorganism and should be present in catalytic amounts.[3] In general, the maximum yield of kinetic resolution is only 50%, which is economically unattractive, but this problem can be overcome for instance by achieving dynamic kinetic resolution in which the unreacted enantiomer is continuously racemised.[6,7]

The most common biocatalysts for the kinetic resolution of racemic compounds is lipases. In the enzyme catalyzed by the transformation of racemic mixture, one enantiomer is readily transferred to the product faster than the other (Fig. 2).[3,8]

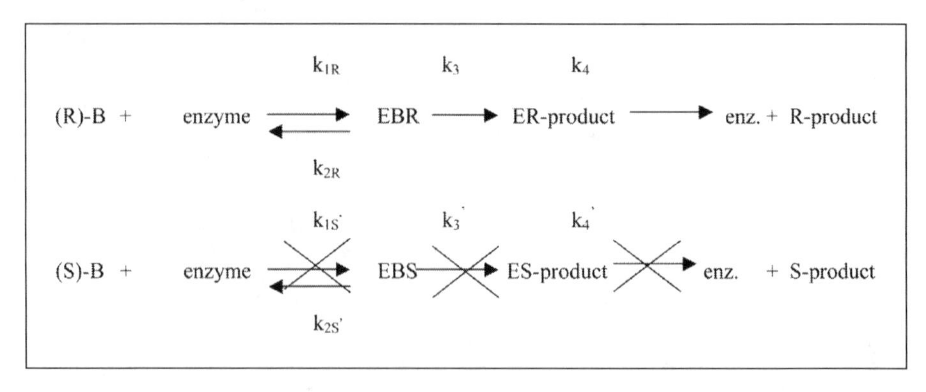

Figure 2. Mechanisms of enzymatic of kinetic resolution of racemate.

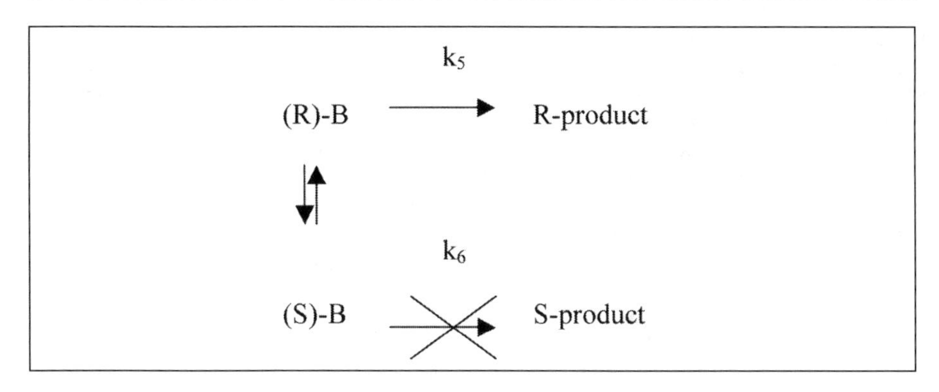

Figure 3. Dynamic kinetic resolution.

(R)-B and (S)-B are the fast- and slow-reacting enantiomers that compete for the same site on the enzyme. For a simple three-step kinetic mechanism, assume the reaction is irreversible. Ideally, 50% conversion of the initial 50:50 mixture occurrs and the final mixture includes 50% unconverted reactant and 50% product (R-product).

Though resolution of two enantiomers provides maximum 50% yield of enantiomerically pure materials at the kinetic resolution, dynamic kinetic resolution gives 100% yield. In the dynamic kinetic resolution, S- enantiomer converts R-enantiomer while R-enantiomer is transferred to the product. Enantiomers are readily separated from each other with enzymes (Fig. 3).[3]

The conversion (x), the enantiomeric excess of the recovered substrate fraction (ee_S), the enantiomeric excess of the recovered product fraction (ee_P), and the value of enantiomeric ratio (E) are calculated by:

$$x = 1 - \frac{(R)-B+(S)-B}{(R)-B_0+(S)-B_0} \tag{1}$$

$$ee_S = \frac{(R)-B-(S)-B}{(R)-B+(S)-B} \qquad \text{for (R) – B > (S) – B} \tag{2}$$

$$ee_P = \frac{R-product - S-product}{R-product + S-product} \qquad \text{for R – product > S – product} \tag{3}$$

$$E = \frac{\ln \frac{[ee_P(1-ee_S)]}{(ee_P+ee_S)}}{\ln \frac{[ee_P(1+ee_S)]}{(ee_P+ee_S)}} \tag{4}$$

The resolution of racemic compounds into individual enantiomer is more economical than asymmetric synthesis. This is because high concentration of substrate can be employed in kinetic resolution.[9]

Chirality is recognized as a key factor in the efficiency of many drug products used to promote human health and agrochemicals used to combate pests which adversely impact the human food supply in use today. The production of single enantiomers of chiral intermediates has become increasingly important, especially in pharmaceutical industry. The close relationship existing between the chemical structure and biological activity of a molecule is a well-verified fact. The absolute configurations of molecular structures determine clear contradictions in the pharmacological activities of many therapeutics groups.[10-12] For many years, synthetic chiral drugs were marketed as racemates. For example ibuprofen was marketed as a non-steroidal anti-inflammatory drug, atenolol as a beta-blocker and albuterol as a bronchodilator. However, this situation is rapidly changing and the trend of enantiopurity in chiral drugs is increasing day by day. Ampicilin as an antibiotic and captopri as an angiotensin-converting enzyme inhibitor are examples of optically pure drugs.[4]

Single enantiomers can be produced by chemical or chemo-enzymatic synthesis.[13] Isolated enzymes and whole-cell enzymes which use chemo-enzmatic synthesis are efficient catalysts under mild conditions.[14] The reactions achieved by biocatalysts are faster and more harmless to the environment. Enzymes act under mild conditions. This minimizes the problem of undesired side-reactions such as decomposition, isomerization, racemization and rearrangement. Biocatalysts can be immobilized and reused for many cycles. Many reactions happening from complex ways or never happening can be achieved easily using biocatalysts. [5,10,15]

Because they are chiral materials and possess distinct substrate selectivity, enzymes are unable to obtain optically active compounds in high enantiomeric excess.[14] They can catalyse numberless reactions that are highyl regio- and enantioselective. Utilization of isolated enzymes and whole-cell enzymes for the synthesis of pharmaceutical products and optically pure drugs as a catalyst has gained substantial importance and searches using biocatalysts have gradually increased.[13,15,16] Active biocatalysts have been obtained by screening a broad variety of microorganism, ranging from archae to fungal cultures.[17] Whole-cell enzymes used as biocatalysts have some advantages to isolated enzymes. While enzyme isolation and purification is very expensive and hard, microbial cells are inexpensive and available in the desired quantities. Despite the success with isolated enzymes as a catalyst for many enantioselective reactions, large quantities of the pure enzymes required for preparative applications are not always readily accessible. Microbial cells are usually used in cofactor-dependent enzymatic reactions in order to allow a cheap regeneration of the cofactor. In contrast to pure enzyme, whole-cells systems (bacteria, fungi) can be available in sufficient quantities.[14,18,19]

Kinetic Resolution of Racemates Using Isolated Enzymes

Lipases are used frequently as chiral catalysts in the synthesis of various fine chemicals and intermediates. The substrate structure and origin of the lipase mainly determine reactivity and selectivity in a lipase-catalyzed reaction. Lipases catalyze the hydrolysis of carboxylic esters and acyl transfer onto hydroxyl and amino groups with the formation of carboxylic esters and amides, respectively. They also catalyze alcohol and acids into the esters. This enzyme accepts a broad range of substrates. Lipases are readily available at low cost and require no cofactor for reaction. Lipases also preserve their activity in the organic medium and have high enzyme activity. The acyl donor used in the acyl transfer reaction acts either as an acid in esterification or an ester in transesterification.

During the transesterification reaction, the lipase is initially acylated by an acyl donor and then acylated lipase reacts enantioselectivitely with an alcohol.[20] The most commonly employed activated esters is vinyl acetate. Several studies have previously shown moderately enantioselective transesterification reactions of racemic alcohol with vinyl acetate.

Lipases can accept a wide range of alcohols as their substrates; however, the enantioselectivity of transesterfication of a racemic alcohol is not always high. In order to increase the enantioselectivity, the effects of changing organic solvents and acyl donors, screening of enzymes and/or temperature, adding an additive, and microwave irridation have been investigated.[2]

Enzyme activity is affected by hydrophobicity of solvent and reactants, solvent polarity and water miscibility. Solvent hydrophobicity and water activity are very important for the retention of enzymatic activity. The solvent hydrophobicity is characterized with log P value. The effects of the solvent on the enzyme activity and enantioselectivity are conflicted. The hydrophobic solvent may deactivate the enzyme in the way of disrupting the functional structure of enzyme or stripping off the essential water from the enzyme.

Suan and Sarmidi[9] investigated the affects of organic solvent with log P values from 1.4 to 4.5 on the resolution of (R,S)-1-phenyl ethanol. They reported that the initial rates increased with enhancement of hydrophibicity of organic solvent. However, the catalytic activity of the enzyme in toluene was lower than the other solvent. They observed that whereas polarity of solvent greatly affected the catalytic activity, it did not affect the enzyme enantioselectivity.

In the another study in the literature, the effect of organic solvent on the kinetic resolution of (R,S)-1-phenyl ethanol with *Mucor mihei* lipase and vinyl propionate as an acyl donor was investigated.[21] MTBE (log P 1.8), hexane /diethyl ether (log P 3.2), hexane (log P 3.5), heptane (log P 4.0), octane (log P 4.5) and decane (log P 5.6) were used as solvents. The activity increased with increasing hydrophibicity of the solvent with exception of hexane. A dependency of enantioselectivity on log P could not be detected.

In the enzymatic reaction, acylated enzyme is the key intermediate. The steric environment of the acyl moiety in the acylated enzyme would be readily modified by employing different acyl donors. It is important to use enantiomerically pure acyl donors. Hirose et al[2] demonstrated the effective kinetic resolution (E = 98) of primary alcohol, 2-pheny-1-propanol, by using a racemic of a chiral cayl donor, vinyl 3-phenyl butanoate.

The anantioselectivity of alcohols has been also increased by using special acyl donors having a phenyl group in the acyl moiety. The enantioselectivity of the lipase from *Pseudomonas cepecia* (PCL) in the transesterification of 2-pheny-1-propanol was investigated using vinyl-3-aryl propionates as acyl donors.[20]

The vinyl-3-(para-substituted phenyl) propionates (R: H, CH_3, C_2H_5, n-C_3H_7, I-C_4H_9, n-C_6H_3O, n-$C_8H_{13}O$, n-$C_8H_{17}O$, C_6H_5, F, Cl, Br, I, CF_3, CN) are more effective acylating agents for the enantioselective transesterification of 2-phenyl-1-propanol compared to the vinyl acetate.

In order to improve the enantioselectivity of the kinetic resolution of the racemic alcohol, another parameter is screening of enzyme. The transesterification of the primary alcohol 2-phenyl-1-propanol have been investigated with various enzymes, which are lipase from *Alcaligenes sp*, *Pseudomonas cepecia*, *Pseudomonas stutzeri* and *Rhizopus sp.*[2] When the chiral acyl donor were used, enantioselectivity of transesterification of 2-phenyl-1-propanol increased with using lipase from *Pseudomonas cepecia*. It is important to increase the substrate concentration for specifically esterification reaction to increase enantioselectivity. Since an esterification reaction catalyzed by a lipase in an organic solvent is reversible, it is common to use an excess of one of the substrate to displace the equilibrium position towards product formation.[22]

The temperature control is accepted as the simple and applicable method for the lipase-catalyzed kinetic resolution of the alcohols. Lipases can prevent their activity in the organic medium. Sakai[23] demonstrated that the low temperature method was easily applied on the kinetic resolution of primer and secondary alcohol. When the solketal shows the low enantioselectivity at 23°C as E 16 and at 0°C as 27 in the lipase catalyzed resolution (*Pseudomonas fluorescens*), its enantioselectivity is increased to 55 at the –40°C temperature. The rate of acceleration is an important subject especially to apply the low-temperature reaction condition. For example, Sakai[23] reported that *Pseudomonas fluerescens* lipase catalyzed resolution of solketal reaction time increased 8-fold compared to those at 30°C. In order to prevent this adversary effect, lipase is immobilized on porous ceramics. Thus aggregation of enzyme is prevented and the reaction rate increased. In the literature, it was also reported that high temperature application increased the enantioselectivity.[24] In this study, the lipase catalyzed transesterification of 1,1-diphenyl-2-propanolin decane, which has a high boiling point (174°C) with excellent enantioselectivity at high temperature up to 120°C in an autoclave using *Pseudomonas cepecia* lipase, immobilized on porous ceramic partricles, toyonite.

Immobilization of an enzyme is known to affect the enzyme conformation, rigidity and reactivity. In an organic solvent, lipase molecules usually form aggregation structures, which reduce the activity. When the lipases are immobilized, they can be highly dispersed.[23]

In the enantioselective esterification of racemic ibuprofen, when the activity of the immobilized lipase was decreased to about 10-20 % that of native lipase, the reaction was more enantioselective with immobilized lipase.[25]

One of the techniques for increasing the enantioselectivityof enzyme is chemical modification. The organic solvent pretreatment for improving the enantioselectivityof lipases has been successful in many cases.[22] *Candida rugosa* lipase (CRL) has been widely used in the resolution of the racemic acids due to high enantioselectivity. However, the commercial CRL has generally been izoenzymes, depending on the commercial lot. In the different percentage polar solvent (methanol, ethanol, acetone, 1-propanol, 2-propanol, 1-bütanol, 2-bütanol), treatment on the crude lipase of CRL were applied by Chamarro et al (2001) and enhanced activity and thermal stability were obtained. Goto et al[27] studied the effects of 2-praponol treatment of the CRL by addition of various volume of 2-propanol. CRL treated with 20 ml 2-propanol has both the highest specific activity and enantioselectivity.

Precipitation of Chromobacterium viscosum lipase by addition of ammonium sulphate and t-butanol, which this method called three phase partition (TPP), led to an increase in the initial rate of transesterification by 4.9 times. TPP appeared to be a suitable strategy for enhancing the overall flexibility of the enzyme in an organic solvent.[28]

The application of microwave on the lipase-catalyzed reaction is the another method for increasing the rates of reaction. Lin and Lin[29] reported that reaction rates and enantioselectivities of the lipase-catalyzed acylation of 1,2,3,4-tetrahydro-1-Naphthol are enhanced 1-14 and 3-9 times, respectively.

Asymmetric Synthesis Using Isolated Enzymes

Asymmetric synthesis has a minor difference from the kinetic resolution. The difference is that the asymmetric atom is not in the molecule to begin with, it is introduced over the course of the reaction.[30] The kinetic resolution of a racemic mixture has an intrinsic limitation in yield: the theoretical maximum yield can not exceed 50% for a single enantiomer.

The microorganisms or isolated enzymes can catalyze transformation of prochiral substrate into a chiral product. The use of commercially available isolated enzymes for the reduction of ketones can provide efficient and economical processes. The production of (S)-3,5-bistrifluoromethylphenyl ethanol by asymmetric enzymatic reduction of a ketone, ADH from *Rhodoccus erythropolis,* have been used. The isolated enzyme ADH reduced ketone with excellent ee (>99.9%) and good conversion (>98%).[31]

Oxidation of alcohols to form carbonyl compounds is one of the most fundamental and important processes in synthetic organic chemistry. Although a variety of methods and reagents have been developed, they all suffer from the general difficulty to transport redox equivalents. Alcohol dehydrogenase (ADH) which oxidize primary and secondary alcohols require $NAD(P)^+$ as a cofactor. $NAD(P)^+$ regeneration techniques must be developed for economic processes (Fig. 4).[32]

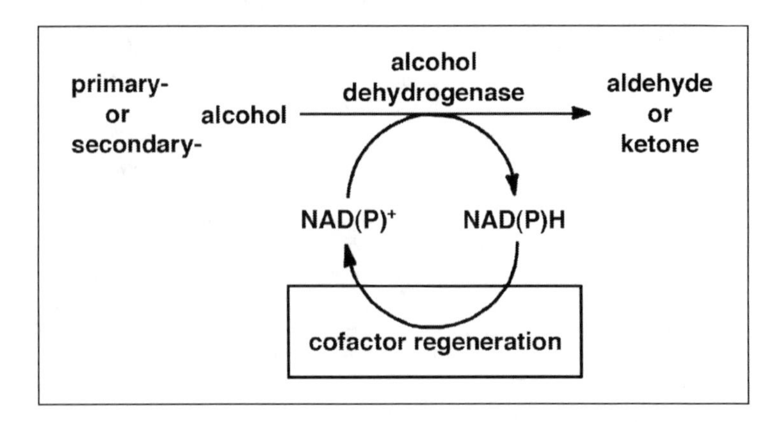

Figure 4. Cofactor recyling for alcohol oxidation employing $NAD(P)^+$ dependent dehydrogenases.

Thamine diphospate (ThDP) enzymes catalyze various lyase and ligase reaction as well as oxidations. With carbo-carbon bond formation, the formation of chiral 2-hydroxyketones, and transferase reactions, new activities have been established for these enzymes. Various ThDP-dependent enzymes use the decarboxylation as a means of substrate activation for a subsequent ligation step. Pyruvate decarboxylase and bezoylformate decarboxylase were already shown to catalyse an acyloin (benzoin)-type condensation reaction, leading to the formation of chiral 2-hydroxyketones.[33]

The combination of enzyme and metal catalysis is also described as a useful method for the synthesis of the optically active compounds. The use of these combinations has been used to catalyze the asymmetric transfer method of ketones, their enol acetates and ketomines. Kim et al[34] have demonstrated that the lipase ruthenium combo catalysis is applicable to the asymmetric reductive transformation of prochiral substrates such as ketones and their enol esters. A broad range of additional enol acetates were transformed into the the corresponding (R)-acetates in good yields with high enantioselectivities.

Kinetic Resolution of Racemates Using Whole-Cell Enzymes

The resolution of racemic compounds into individual enantiomers can be achived succesfully by microbial cells (bacteria and fungus). In addition to isolated enzyme and desired enantiomer, you can find good excess and good conversion. Whole-cell enzymes using as biocatalyst have some advantages to isolated enzymes. While enzyme isolation and purification is very expensive and hard, microbial cells are inexpensive and available in the desired quantities. This section gives examples using microbial cells-catalyzed kinetic resolution for the synthesis of single enantiomers of key intermediates for drug substances.

The microbial cells-catalyzed resolution of racemic mixture of secondary alcohol (2-octanol, 2-pentanol, 2-hekzanol, 2-heptanol) by direct esterification in organic solvents were investigated by Molinari et al.[35] Chiral secondary alcohols are a common sub-structure within many pharmaceutical drug candidates; for instance, (S)-2-pentanol and (S)-heptanol are intermediates in the synthesis of several potential anti-Alzheimer's drugs that inhibits synthesis and release of β-amyloid peptide.[36] Molinari et al[35] have previously observed that liyophilized cells of *M. letus* ATCC 9341, *R. delemar* MIM and *R. oryzae* CBS 112-07 are able to catalyze the esterification of racemic 2-octanol with butyric acid in heptane. In this work they used *M. letus* ATCC 9341, *R. delemar* MIM and *R. oryzae* CBS 112-07 as liyophilized cells. For the esterification of racemic 2-octanol with butyric acid, *Rhizopus oryzae* CBS 112-07 was selected as a suitable biocatalyst and >97% enantiomeric excess of (R)-ester and 41 % molar conversion was found as the highest result. Esterification of the other racemic alcohol showed that using short-chain alcohol decreased the enantiomeric excess.

Snell et al[37] investigated enantioselective hydrolysis of racemic ibuprofen amide to S-ibuprofen [(R,S)-2- (4-isobutylphenyl) propanoic acid] which belongs to a group of potent, orally active, nonsteroidal, anti-inflammatory agents by *Rhododococcus* AJ270 and at the end biologically active S-ibuprofen was synthesied in very high enantiomeric excess (90-94%).

The microbial direct esterification of racemic 2-phenyl-1-propanoic acid with etanol by liyophilized mycelium of *Aspergillus oryzae* MIM and *Rhizopus oryzae* CBS 112.07 was reported by Gandolfi et al.[38] It was found that *R. oryzae* CBS 112.07 gave (R)-ethyl ester with 97 % enantiomeric excess and *A. oryzae* MIM (S)-ethyl ester with 90 % enantiomeric excess.

The synthesis of enantiomerically enriched 2-aryl-4-pentenoic acids derivatives (3) and 2-aryl-4-pentenamides (2) by microbial hydrolysis of nitriles (1) was studied by Wang et al[39] (Fig. 5). They had shown in pervious studies that kinetic resolution of racemic trans-2-arylcyclopropanecarbonitriles were affected efficiently by microbial *Rhododococcus* sp. AJ270 cells to produce enantioselective synthesis of 2-arylcyclopropanecarboxylic acids and 2-arylcyclopropanecarboxamides in high enantiomeric excess.[40] In this work, *Rhododococcus* sp. AJ270 was used as a whole-cell catalyst and almost all of the substrate investigated gave good enantiomeric excess for (S)-acids (87.4 to >99.5%) and (R)-amides (99.2 to >99.5%).

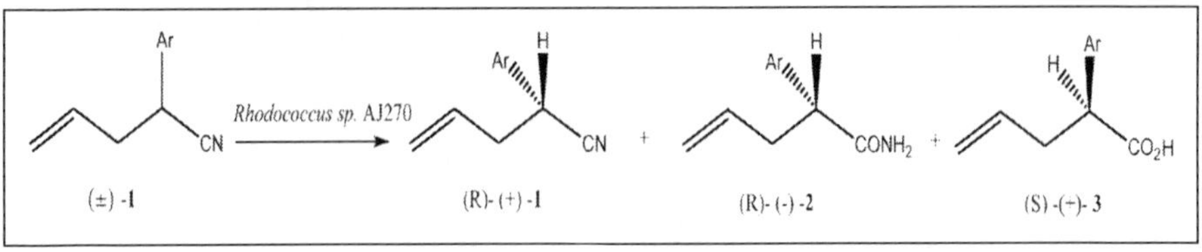

Figure 5. Hydrolysis of nitriles (±)-1.

Benzoin-type chiral α-hydroxy ketones are important classes of intermediates in organic synthesis due to their biofunctional aspect, especially having one chiral center amenable to further modification.[41] They are important structural units in many biologically active natural products and recently they have been used as convenient building blocks in organic synthesis. Chiral α-hydroxy ketones can be prepared enzymatically by kinetic resolution of the racemate of either α-peroxo-, α-hydroxy or α-acetoxy ketones.[42-45] Demir et al[42] studied enantio selective synthesis of α-hydroxy ketones via hydrolysis of α-acetoxy aryl alkyl ketones using the fungus *Rhizopus oryzae*. Their results show that *Rhizopus oryzae*-catalyzed hydrolysis of α-acetoxy ketones provides α-hydroxy ketones with high enantiomeric excess and enantiomerically enriched α-acetoxy ketones in good chemical yields. For R-form of benzoin, ee and yield was found respectively as 36 % and 70%. In the other study, chemoenzymatic synthesis of pharmacologically interesting (R)-2-hydroxypropio phenones starting from propiophenone derivatives was performed by fungus-mediated hydrolysis of the acetoxy group.[46] The results show that the enzymes found in the *Rhizopus oryzae* fungus favored the (R)-enantiomers in this conversion reactions and (R)-2-hydroxypropiophenones were obtained with high enantiomeric excesses (92-97%) and in good yields (46-34%). Babaarslan et al[47] investigated the preparation of enantioselective benzoin, which is one of the α-hydroxy ketone by kinetic resolution of racemic benzoin, with liyophilized microorganisms (bacteria and fungi). They used liyophilized *pseudomonas fluorescens* Biovar I, *Pseudomonas fluorescens, Pseudomonas putida, Pseudomonas stutzeri, Pseudomonas aeroginase, Rhizopus oryzae* CBS 112-07 and *Aspergillus oryzae* as a lipase source in kinetic resolution of racemic benzoin via transesterication (Fig. 6) and investigated affect of acyl donor type, solvent type, lyophilized microorganism amount, moeculer sieves amount and temperature in the kinetic resolution of racemic benzoin.

They report that these microbial lipases could not discriminate between the two enantiomers of benzoin and both enantiomers were transferred to the product at nearly same rate. The investigated parameters affected the conversion of rasemic benzoin, but enantiomeric excess of racemic benzoin did not change.

Matsumata et al[48] researched the preparation of optically active C_2-symmetrical diols, which are useful as chiral auxilaries and used as chiral catalyst in asymmetric reactions via the microbial hydrolysis of the racemic cyclic carbonates (Fig. 7). In this work, *pseudomonas dimunuta* was selected as the best strain to perform the stereoselective hydrolysis. Moreover, it was found that the ring size did not affect the reactivity and enantioselectivity, and a six-membered cyclic carbonate, *dl*-4,6-dimethyl-1,3-dioxan-2-one, was easily hydrolyzed with higher enantioselectivity to afford the optically active (S,S).

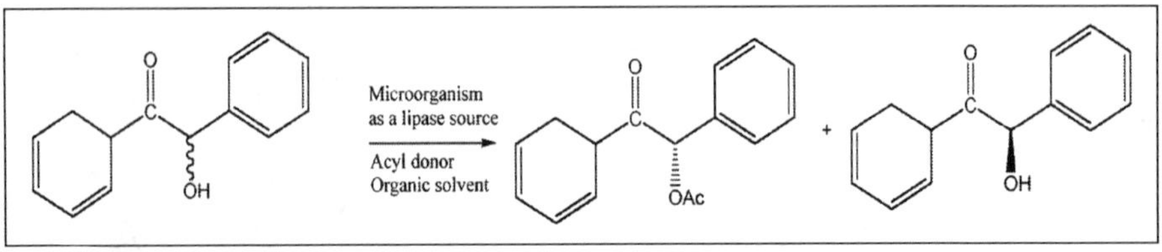

Figure 6. Kinetic resolution of racemic benzoin via transesterification.

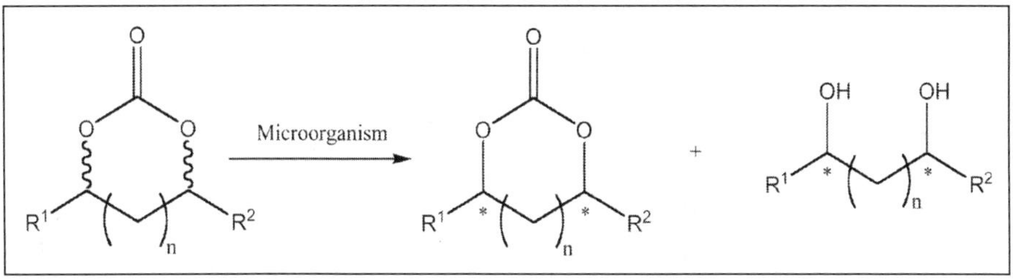

Figure 7. The microbial hydrolysis of the racemic cyclic carbonates.

Hydrolytic kinetic resolution of racemic 1-arylethyl acetates by *Pseudomonas fluorescens* RRLJ 134 in the presence of a surfactant was studied by Bora et al.[19] They showed that the *P. fluorescens* RRLJ 134 might be used as a whole cell—catalyzed in the hydrolysis of acetates into symmetrical alcohols. They reported the use of *P. fluorescens* in the hydrolysis of racemic 1-arylethyl acetates with a slight preference for the S- alcohol. Moreover, they found that the enantioselectivity of the biocatalyst could have been completely reversed and significiantly enhanced by using the surfactant Tween 80.

Asymmetric Synthesis Using Whole Cell Enzymes

Microbial cells are usually used in cofactor-dependent enzymatic reactions in order to allow a cheap regeneration of the cofactor. Optically pure materials can be synthesied by way of transformation of prochiral substrate with asymmetric synthesis. This section gives examples of microbial cells-catalyzed asymmetric synthesis for the synthesis of single enantiomers of key intermediates for drug substances.

Asymmetric reduction of prochiral ketones is one of the most investigated methods to produce optically active compounds.[49] Microbial reduction of prochiral ketones have several advantages: regio- and enantioselectivity, in vivo cofactor regenaration, mild reaction conditions and enviromentally friendly operation.[50]

The preparation of halohydrin using yeast-catalyzed reduction of α-haloketones were investigated by Martinez et al[51] (Fig 8). Chiral halohydrins are valuable synthetic intermediates for the preparation of a wide range of biologically interesting compounds and are widely used as building blocks to prepare β-blockers.[45,52] They have performed the stereoselective reduction of 1-aryloxy-3-halopropan-2-ones (**1**) to 1-aryloxy-3-halopropan-2-ol (**2**) 1 with good yield and ee using different microorganism strains and experimental conditions. They found that *Yarrowia lipolytica 1240* resting cells gave 87 % yield of (S)-form (99% ee) and *Pichia mexicana* 11105 resting cells gave 85% yield of (R)-form (95% ee) for 1-chloro-3-(1-naphthloxy) propan-2-ol.

Martinez Lagos et al[53] researched the stereoselective reduction of the same α-haloketones to the halohydrin with three new yeasts. They have selected the new yeasts, *S. bayanus* CECT 1317, *Y. lipolytica* CECT 1240 and *P. mexicana* CECT 1015, after a taxonomical screening for microorganisms active in the reduction of ketones. It was found that *P. mexicana* gave ee greater than

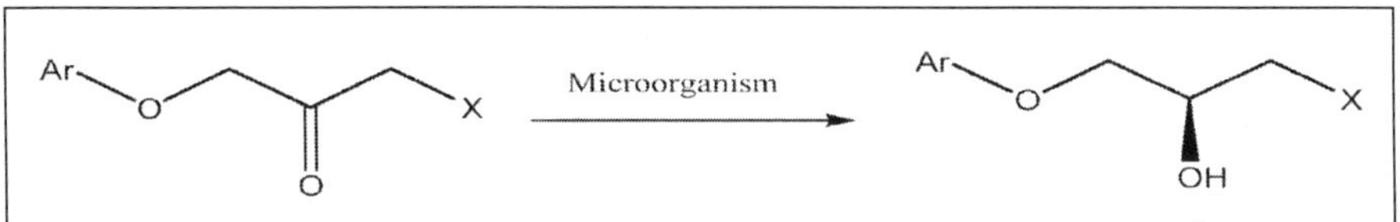

Figure 8. Biotransformation of 1-aryloxy-3-halopropan-2-ones.

90% and yields higher than 85% for the (S)- or (R)-halohydrins. In the following studies, they perfomed reduction of 1-chloro-3-(1-naphthyoxy)-2-propanone by immobilized yeasts which were *P. mexicana* and *Y. lipolytica*. They observed that the immobilization slightly reduced the enzymatic activity in the case of *P. mexicana* compared to the resting cells, whereas *Y. lipolytica* maintained the activity. The yield of halohydrin was found to be 80% for R-form (96% ee) with *P. mexicana* and 80 % for S-form (95-97%) with *Y. lipolytica*.[52]

Erdelyi et al[50] reported the bioreduction of different prochiral aryl- and aralkyl- ketones to their corresponding alcohols, which are possible intermediates in the synthesis of further drug candidates. The yeast strain *Z. rouxii* and *D. hanseii* were choosen as the best biocatalysts between investigated yeasts respective to enzyme activity and enantioselectivity. It was found that *Z. Rouxii* ATCC 14462 produced (S)-alcohols 1-phenylpropan-2-ol, 1-(4-metoxyphenyl)propan-2-ol, 1-(4-chlorophenyl) propan-2-ol in good yields and with excellent enantioselectivies. The ortho-substituted phenylacetones, benzylacetones and acetophenones were reduced in sufficent yields and enantiomeric purities with *D. hanseii*.

Asymmetric synthesis of chiral α-hydroxy ketones, which are important structural units in many biologically active natural products, prepared enzymatically by reduction of the α-diketones with baker's yeast[54] and asymmetric acyloin condensation.[41,46,55] Nakamura et al[54] achieved enantio- and regioselective reduction of α-diketones with baker's yeast. Their results showed that yeast-catalyzed reduction of α-diketones afforded a mixture of two α-hydoxy ketones and *vic*-diol. The use of an enzyme inhibitor prevented the production of the diol and regioselectivity in the reduction to α-hydroxy ketone was improved by thermal pre-treatment of baker's yeast. In conclusion, for 1-phenyl-2-hydroxy-1-propanone ee and yield was found respectively as >98% and 80%. Fungi-mediated conversion of benzil to benzoin and hydrobenzoin was researched by Demir et al[56] (Fig. 9). They found that depending on the pH of the medium, both enantiomers of benzoin could be prepared in good yield and high ee and the reduction of benzoin isomers furnished (R,R)-hydrobenzoin in high ee and good yield. They used the growing cells of *Rhizopus oryzae* fungus and found the best result for R-form of benzoin with *Rhizopus oryzae* ATCC 9363 as >99 (ee) at pH 6.5-8.5 and for S-fom of benzoin with *Rhizopus oryzae* 72465 as 85% (ee).

Babaarslan et al[57] investigated enantioselective reduction of benzil to bezoin using *R. oryzae* CBS 111718 and *R. oryzae* CBS 112-07 by growing cells and resting cells at different pH values. Using growing and resting cells of *R. oryzae* CBS 111718 and *R. oryzae* CBS 112-07 obtained both enantiomers of benzoin, depending on the pH of the medium. The highest ee was found as 87% for S-form of benzoin with growing cells of *R. oryzae* at pH :7.8, and 76% for R-form of benzoin at pH:6.2.

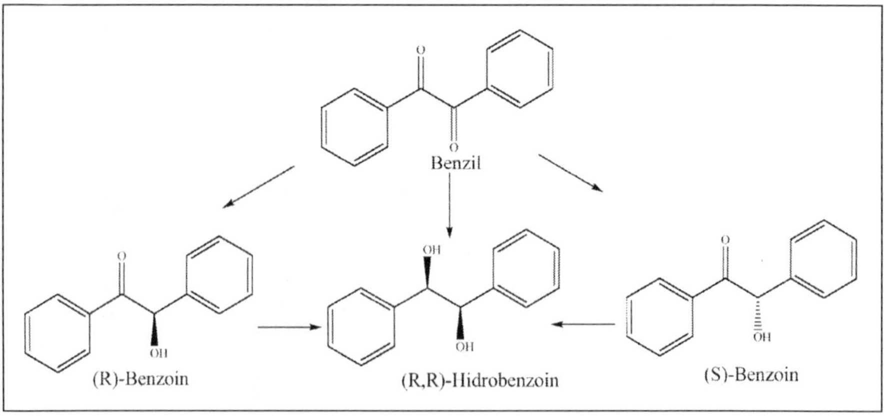

Figure 9. Enantioselective reduction of benzil to benzoin.

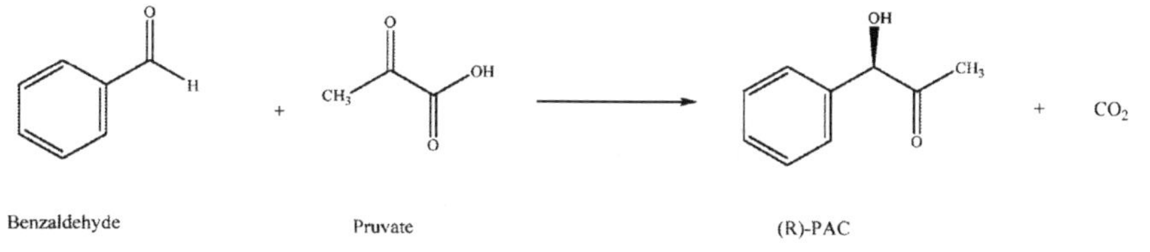

Figure 10. Biotransformation of benzaldehyde and pruvate into (R)-PAC.

L-Phenylacetylcarbinol (L-PAC), also known as 1-hydroxy-1-phenyl-2-propanone, can be given as an example of a microbial condensation reaction. L-PAC is an intermediate for commercial chemical synthesis of L-ephedrine and L-pseudoephedrine, an ingredient of pharmaceutical preparations used as anti-asthmatics and decongestants, and is currently produced industrially via a biotransformation of benzaldehyde (Fig. 10) by fermenting yeast cultures.[58,59]

The production of PAC was performed by growing cells of yeast, resting cells and immobilizing whole cells of *Saccharomyces cerevisiae* and *Zygosaccharomyces roixii* in a growth medium.[58] Strains belonging to the genera *Saccharomyces* and *Candida* have been observed to be more efficient L-PAC producers in comparison to the other yeasts by Tripathi et al.[60] A final concentration of L-PAC was found as 15.2 g/L with immobillized *Candida utilis* in a fed-batch process[61] and as 29 g/L in a stirred-tank reactor with immobilized *Saccharomyces cerevisiae.* [58]

Olivo et al[62] reported microbial oxidation/amidation of benzhydrylsulfanyl acetic acid for synthesis of (+)-modafinil. Modafinil is a psychostimulant agent that has gained a lot of attention because of its recent approval by the FDA for treatment of excessive daytime sleepiness and because of its lack of abuse liability. Modafinil might also be of utility as a treatment of attention deficit/hyperactivity disorder and in treating opioid-induced sedation. In this work, a highyl enantioselective oxidation of benzhydrylsulfanyl acetic acid to (S)-sulfinyl carboxylic acid was perfomed using the fungus *Beauveria bassiana* in very good yield (ee: 99%, yield: 89%). This product was amidated with the bacteria *Bacillus subtilis* to (S)-modafinil in good yield (ee: 68%, yield: 100%).

References

1. Maier MN, Franco P, Lindner W. Separation of enantiomers: needs, challenges, perspectives. J Chomotogr A 2001; 906:3-33.
2. Hirose K, Naka H, Yano M et al. Improvement of enantioselectivity in kinetics resolution of a primary alcohol through lipase-catalysed transesterification by using a chiral acyl donor. Tetrahedron: Asymmetry 2000; 11:1199-1210.
3. Ghanem A, Aboul-Enein HY. Lipase-mediated chiral resolution of racemates in organic solvents. Tetrahedron: Asymmetry 2004; 15:3331-3351.
4. Sheldon RA. Synthetic Methodology. Chirotechnology. New York; Marcel Dekker, 1993:49-129.
5. Kurt F. Biocatalytic Applications. Biotransformation in Organic Chemistry. 4th ed., Berlin: Springer, 2000:29-332.
6. Straathof AJJ, Panke S, Schmid A. The production of fine chemicals by biotransformation. Curr Opin Biotechnol 2002; 13:548-556.
7. Sundbay E, Andersen MM, Hoff BH et al. Lipase catalysed resolution and microbial reduction for obtaining enantiopure 1-(2-thienyl) alkonols. Arkivoc 2001; 76-84.
8. Chen C, Fujimoto Y, Girdaukas G et al. Quantitave analyses of biochemical kinetic resolutions of enantiomers. J Am Chem Soc 1982; 104:7294-7299.
9. Suan C, Sarmidi MR. Immobilized lipase-catalysed resolution of (R,S)-1-phenyl ethanol in recirculated packed bed reactor. J Mol Catal B:Enzymatic 2004; 28:111-119.
10. Patel RN. Enzymatic synthesis of chiral intermeadiates for drug development. Adv Synth Catal 2001; 343:527-546.
11. Crosby J. Synthesis of optically active compounds. A large scale perspective. Tetrahedron 1991; 47:4789-4846.
12. Bermudez JL, Campo C, Salazar L et al. A new application of Candida anthartica lipase for obtaining natural homochiral BBAs aryloxypropanolamine. Tetrahedron: Asymmetry 1996; 7:2485-2488.

13. Patel RN. Microbial/enzymatic synthesis of chiral intermediates for pharmaceuticals. Enzyme Microb Technol 2002; 31:804-826.
14. Adam W, Heckel F, Saha-Möller CR, Schreier B. Biocatalytic synthesis of optically active oxyfunctionalized building blocks with enzymes, chemoenzymes and microorganism. J Organametal Chem 2002; 661:17-29.
15. Wu JY, Liang MT. Enhancement of enantioselectivity by altering alcohol concentration for esterification in supercritical CO_2. J Chem Eng Japan 1999; 32:338-340.
16. Leuenberger HGW. Biotransformation- a useful tool in organic chemistry. Pure & Appl Chem 1990; 62:753-768.
17. Schoemaker HE, Mink D, Wubbolts MG. Dispelling the myths-biocatalysis in Industrial synthesis. Science 2003; 299:1694-1697.
18. Adam W, Lukacs Z, Saha-Möller CR et al. Biocatalytic kinetic resolution of racemic hydroperixides through the enantioselective reduction with free and immobilized. J Am Chem Soc 2000; 122:4887-4892.
19. Bora U, Saikia CJ, Chetia A et al. Resolution of racemic 1-arylethyl acetates by Pseudomonas fluorescens in the presence of a surfactant. Tetrahedron Lett 2003; 44:9099-9102.
20. Kawasaki M, Goto M, Kawabata S et al. The effect of vinyl esters on the enantioselectivity of the lipase-catalysed transesterification of alcohols. Tetrahedron: Asymmetry 2001; 12:585-596.
21. Frings K, Koch M, Hartmeier W. Kinetics Resolution Of 1-phenyl ethanol with high enantioselectivity with native and immobilized lipase in organic solvents. Enzyme Microb Technol 1999; 25:303-309.
22. Berglund P. Controlling lipase enantioselectivity for organic synthesis. Biomol Eng 2001; 18:13-22
23. Sakai T. Low-temperature method' for a dramatic improvement in enantioselectivity in lipase-catalyzed reactions. Tetrahedron: Asymmetry 2004; 15:2749-2756.
24. Ema T. Rational strategies for highly enantioselective lipase-catalyzed kinetic resolutions of very bulky chiral compounds:substrate design and high-temperature biocatalysis. Tetrahedron: Asymmetry 2004; 15:2765-2770.
25. Ikeda Y, Kurokawa Y. Enantioselective esterification of racemic ibuprofen in isooctane by immobilized lipase on cellulose acetate-titanium iso-propoxide gel fiber. J Bioscience Bioeng 2002; 93(1):98-100.
26. Chamorro S, Alcantara A R, Casa R M et al. Small water increase the catalytic behaviour of polar organic solvents pre-treated Candida rugosa lipase. J Mol Catal B: Enzymatic 2001; 11:939-947.
27. Goto M, Ogawa H, Isobe T et al. 2-propanol teratment of Candida rugosa lipaseand its hydrolytic activity. J Mol Catal B: Enzymatic 2003; 24-25:67-73.
28. Roy I, Gupta M N. Enhancing reaction rate for transesterification reaction catayzed by Chromobacterium lipase. Enzyme Microb Technol 2005; 36:869-899.
29. Lin G, Lin W. Microwave-Promoted Lipase-Catalyzed Reactions. Tetrahedron Lett 1998; 33:4333-4336.
30. Eliel E L, Racemic Modifications. In: Fernelius WC, ed. Stereochemistry of Carbon Compounds., Tokyo: McGraw-Hill Book Company, 1962:31-85.
31. Pollard D, Truppo M, Pollard J et al. Effective synthesis of (S)-3,5-bistrifluoromethylphenyl ethanol by asymmetric enzymatic reduction. Tetrahedron: Asymmetry 2006; in press.
32. Kroutil W, Mang H, Edegger K et al. Biocatalytic oxidation of primary and secondary alcohols. Adv Synth Catal 2004; 346:125-142.
33. Pohl M, Sprenger GA, Müler M. A new perspective on thiamine catalysis. Curr Opin Biotechnol 2004; 15:335-342.
34. Kim M J, Ahn Y, Park J. Dynamic kinetic resolution and asymmetric transformations by enzymes coupled with metal catalysis. Chem Biotechnol 2002; 13:578-587.
35. Molinari F, Mantegazza L, Villa R et al. Resolution of 2-alkonols by microbially-catalyzed esterification. J Ferment Bioeng 1998; 86:62-64.
36. Patel RN. Biocatalytic synthesis of chiral pharmaceutical intermediates. Food Technol Biotechnol 2004; 42:305-325.
37. Snell D, Colby J. Enantioselective hydrolysis of racemic ibuprofen amide to s-(+)-ibuprofen by Rhodoccoccus AJ270. Enzyme Microb Technol 1999; 24:160-163.
38. Gondolfi R, Gualandris R, Zanchi C et al. Resolution of (RS)-2-phenylpropanoic acid by enantioselective esterification with dry microbial cells in organic solvents, Tetrahedron: Asymmetry 2001; 12:501-504
39. Wang MX, Zhao SM. Synthesis of enantiomerically enriched (S)-(+)-2-aryl-4- pentenamides via microbial hydrolysis of nitriles, a chemoenzymatic approach to stereoisomers of α,γ-disubstituted γ-butyrolactons. Tetrahedron: Asymmetry 2002; 18:1695-1702.
40. Wang MX, Feng GQ. Enantioselective synthesis of chiral cyclopropane compounds through microbial transformations of trans- 2-arylcyclopropanecarbonitriles. Tetrahedron Lett 2000; 41:6501-6505.
41. Guo Z, Goswami A, Mirfakhrae KD. Aymmetric acyloin condensation catalyzed by phenylpyruvate decarboxylase. Tetrahedron: Asymmetry 1999; 10:4667-4675.

42. Demir AS, Hamamci H, Tanyeli C et al. Synthesis and Rhizopus oryzae mediated enantioselective hydrolysis of α-acetoxy aryl alkyl ketones. Tetrahedron Lett 1998; 9:1673-1677.
43. Kajiro H, Mitamura S, Mori A et al. Enantioselective synthesis of 2-hydroxy-1-indanone, a key precursor of enantimerically pure 1-amino-2-indanol. Tetrahedron: Asymmetry 1998; 9:907-910.
44. Adam W, Diaz MT, Fell RT et al. Kinetic resolution of racemic α-hydroxy ketones by lipase- catalyzed irreversible transesterification. Tedrahedron: Asymmetry 1996; 7:2207-2210.
45. Zhu D, Mukherjee C, Hua L. Green synthesis of important pharmaceutical building blocks: enzymatic access to enantiomerically pure α-chloroalcohols. Tetrahedron: Asymmetry 2005; 16:3275-3278.
46. Demir AS, Pohl M, Janzen E. et al. Enantioselective synthesis of hydroxy ketones through cleavage and formation of acyloin linkage. Enzymatic kinetic resolution via C-C bond cleavage. Perkin Trans 2001; 1344:633-635.
47. Babaarslan C, Mehmetoglu U, Demir AS. Kinetic Resolution of Racemic Benzoin with Different Lyophilized Microorganisms. J Biotechnol 2005; 118:49.
48. Matsumato MK, Shimojo Y, Hatanaka M. Highly enantioselective preparation of C2-symmetrical diols: microbial hydrolysis of cyclic carbonates. Tetrahedron: Asymmetry 2000; 11:1965-1973.
49. Ribeiro JB, Ramos KV, Neto FR et al. Microbiological enantioselective reduction of ethyl acetoacetate. J Mol Catal B: Enzymatic 2003; 24-25:121-124.
50. Erdelyi B, Szabo An, Seres G et al. Stereoselective production of (S)-1-aralkyl- and 1-arylethanaols by freshly harvested and lyophilized yeast cells. Tetrahedron: Asymmetri 2006; 17:268-274.
51. Martinez F, Campo CD, Sinisterra JV et al. Preparation of halohydrin β-blockers using yeast-catalysed reduction. Tetrahedron: Asymmetry 2000; 11:4651-4660.
52. Martinez-Lagos F, Sinisterra JV. Enantioselective production of haohydrin precursor of propranolol catalysed by immobilized yeasts. J Mol Catal B: Enzymatic 2005;36:1-7.
53. Lagos FM, Carballeira JD, Bermudez JL et al. Highly stereoselective reduction of haloketones using three new yeasts: application to the synthesis of (S)- adrenergic β-blockers related to propranolol. Tetrahedron: Asymmetry 2004;15 :763-770.
54. Nakamura K, Kondo Sİ, Kawai Y et al. Enantio- and Regioselective reduction of α-diketones by Baker's yeast. Tetrahedron: Asymmetry 1996;7:409-412.
55. Demir AS, Şeşenoğlu Ö, Eren E et al. Enantioselective Synthesis of -Hydroxy Ketones via Benzaldehyde Lyase-Catalyzed C-C Bond Formation Reaction. Adv Synth Catal 2002; 344:96-103.
56. Demir AS, Hamamcı H, Ayhan P et al. Fungi mediated conversion of benzil to benzoin and hydrobenzoin. Tetrahedron: Asymmetry 2004; 15:2579-2582.
57. Babaarslan Ç, Mehmetoglu Ü, Demir AS. Rhizopus Oryzae CBS 111718 Catalysed Enantioselective Benzoin Synthesis via Reduction of Benzil. Multistep Enzyme Catalysed Processes Conference 2006: In press, Graz, Austrian.
58. Mandwal AK, Tripathi CKM, Trivedi PD et al. Production of ι-phenylacetyl carbinol by immobilized cells of Saccharomyces cerevisiae. Biotechnol Lett 2004; 26:217-221.
59. Rosche B, Sandford V, Breuer M et al. Biotransformaion of benzaldehyde into (R)-phenylacetylcarbinol by filementous fungi or their extracts. Appl Microbiol Biotechnol 2000; 57:309-315.
60. Tripathi CM, Agarwal SC, Basu SK. Production of L-phenylacetylcarbinol by fermentation J Ferment Bioeng 1997; 84:487-492.
61. Shin HS, Rogers PL. Biotransformation of benzeldehyde to l-phenylacetylcarbinol. an intermediate in l-ephedrine production, by immobilized Candida utilis. Appl Microbiol Biotechnol 1995; 44:7-14.
62. Olivo HF, Osorio-Lozada A, Peeples TL. Microbial oxydation/amidation of benzhydrylsulfanyl acetic acid. Synthesis of (+)-modafinil. Tetrahedron: Asymmetry 2005; 16:3507-3511.

CHAPTER 8

Enzyme Catalysis in Fine Chemical and Pharmaceutical Industries

Ganapati D. Yadav,* Ashwini D. Sajgure and Shrikant B. Dhoot

Introduction

Biotechnology is being increasingly adopted by several chemical companies to improve manufacturing sustainability and profitability as regards energy consumption and feed stock access as well as the production of high-value chemicals.[1,2] Industrial biotranformations generally center around natural compounds such as carbohydrates and fats in food sector (Fig. 1).[3] However, the pharmaceuticals sector has started dominance in this area (Fig. 2).[4] Chemical and pharmaceutical industries need eco-friendly and benign technologies and thus waste minimization of both material and energy and catalysis are at the hub of all green processes. A practical alternative to the traditional organic synthesis is biotechnology and in particular enzymatic catalysis is weighed against chemical catalysis in fine chemical and pharmaceutical industries. Enzymes are exquisitely selective in catalyzing reactions with unparalleled chiral and regio selectivities. Enzymes typically operate at room temperature and ambient pressure and lead to a few waste products; neither do they need the tedious blocking and deblocking steps that are common in enantio and regio selective organic synthesis. Chirality is of utmost importance in drug synthesis. Chiral nature of the enzymes is responsible for their chemo, regio- and stereospecificity. A typical enzyme with a mass of 50,000 daltons possesses 450 amino acid residues: 19 chiral L-aminoacids and glycine and further if glycine makes up 10% of the residues, then there are at least 400 residues with chiral centers to provide an asymmetric environment for substrate binding and subsequent chemical transformation. There is a requirement to develop processes for the production of enantiomerically pure drugs. The market for single isomers of chiral drugs is more than 100 billion US dollars and continue to expand rapidly.[5]

Enzymes function not only in aqueous media but also in organic solvents, which has led to their versatility for applications in fine chemical and drug synthesis. Many of them are active in pure solvents or in supercritical fluids in the absence of added water. Use of organic solvents in enzymatic reactions offers various advantages such as higher substrate solubility, reversal of hydrolytic reactions and modified enzyme specificity, which result in better or novel enzyme activities that previously were only possible by using genetic modifications or complex multistep pathways within whole cells. Enzymes are able to dissolve in hydrophobic organic solvents, if low concentration of a suitable surfactant is present. They remain remarkably active and with secondary and tertiary structures, their activity is nearly identical to that measured in water.[7]

*Corresponding Author: Ganapati D. Yadav—Department of Chemical Engineering, University Institute of Chemical Technology, University of Mumbai, Matunga, Mumbai 400019, India. Email: gdyadav@udct.org, gdyadav@yahoo.com

Enzyme Mixtures and Complex Biosynthesis, edited by Sanjoy K. Bhattacharya.

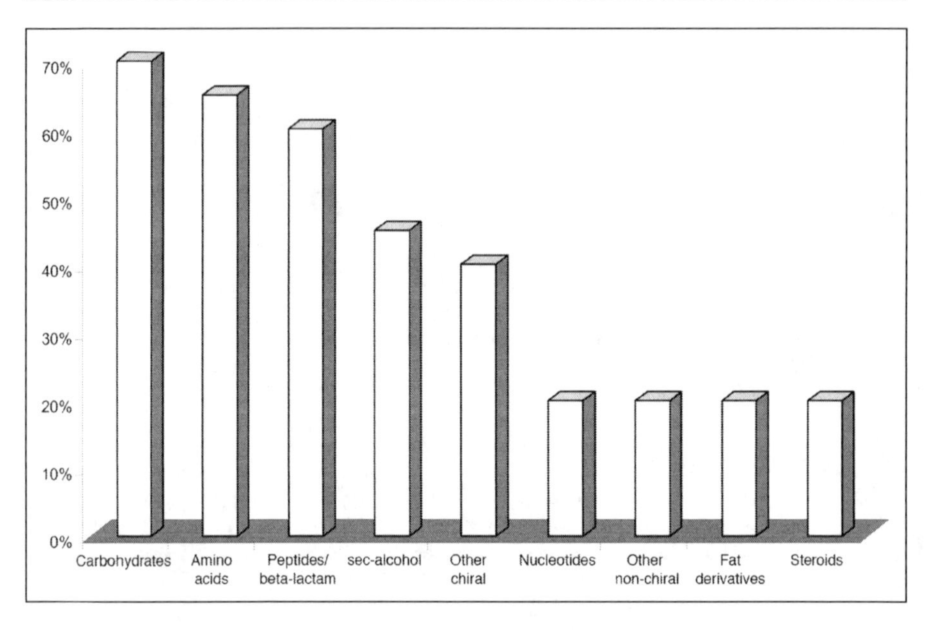

Figure 1. Classes of compound produced by biotransformation (adapted from ref. 4).

There is an upsurge in the use of biocatalysts to prepare a range of products for the pharmaceutical industry (e.g., 6-aminopenicillanic acid), food and nutrition (e.g., glucose and fructose syrups, l-lysine and niacinamide), as well as speciality and commodity chemicals (e.g., acrylamide and acrylic acid).[8] The importance of integrated process development involving chemo-enzymatic synthesis is realized for success of industrial biocatalysis. Apart from the traditional lipase catalyzed reactions, new reactions evaluated in industry now include oxidations[9] and carbon-carbon bond formations.[10] This Chapter covers enzyme catalysis in fine chemical and pharmaceutical synthesis. The trends in some bulk chemical preparations are also given as a prelude.

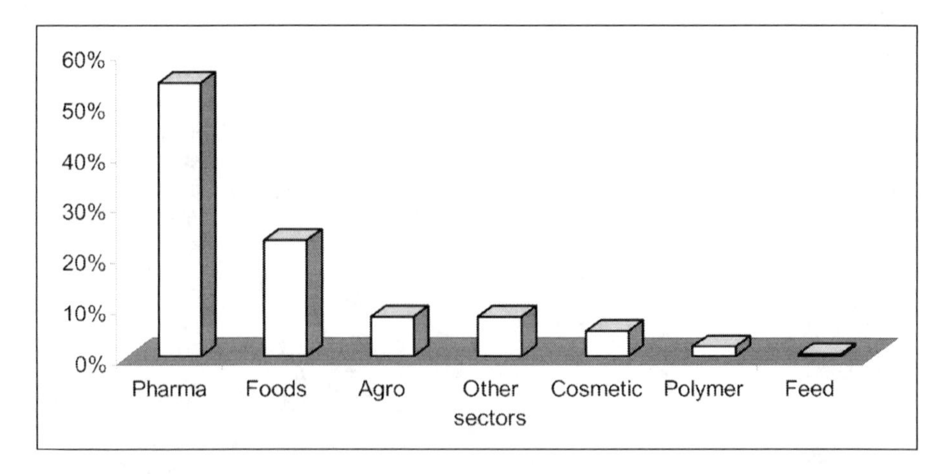

Figure 2. Applications of biotranformations in various industrial sectors (adapted from ref. 4).

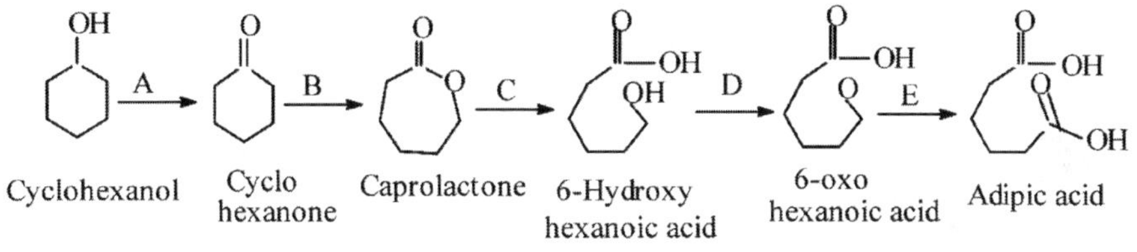

Cyclohexanol → A → Cyclo hexanone → B → Caprolactone → C → 6-Hydroxy hexanoic acid → D → 6-oxo hexanoic acid → E → Adipic acid

Figure 3. Enzymatic oxidation of cyclohexanol to adipic acid (A. cyclohexanol dehydrogenase, B. cyclohexanone monooxygenase, C. caprolactone esterase, D. 6-hydroxy hexanoate dehydrogenase, E. 6-oxohexanoate oxidoreductase).[12]

Biotransformations in Bulk and Fine Chemicals

Cycloalkanone to Carboxylic Acid

Caprolactone, the largest volume lactone with a global production of ~150,000 TPA, is produced from cyclohexanone and peracetic acid by using Baeyer-Villiger oxidation at 50°C with selectivity of 85-90%.[11] In the alternate biochemical route, cyclohexanone and cyclohexanol are metabolized through a series of sequential oxidation reactions involving several enzymes (Fig. 3). A large number of microorganisms can oxidize cyclohexanol and cyclohexanone to adipic acid.[12]

Cyclohexanone oxidation to caprolactone is catalyzed by a Baeyer-Villiger monooxygenase,[13] which is also used to produce a wide range of compounds.[14] Bulky cyclic ketones can be oxidized biochemically by two *Rhodococcus sp.*[15,16] For instance, cyclodecanone is converted to dodecanadoic acid in an analogous pathway. Baeyer-Villiger monooxygenase from *Acinetobacter sp.* NCIMB 9871 is used mostly in the reactions.[15,17]

Alkylaromatic Oxidation

Oxidized alkylaromatics find applications in the manufacture of films, fibers, paints, adhesives and softeners. In the production of oxidized alkylaromatics, enzymes are used in the initial purification of xylenes and ethylbenzene mixture and in the generation of partially oxidized intermediates. Enzymes that have specificity toward either *p*-xylene or *o*-xylene are used to simplify the purification step. Several genes have been reported that encode enzymes that can oxidize alkyl groups attached to aromatic rings.[18]

Nitrile Bioconversions

Microbial catalysts with nitrile hydratase or nitrilase enzyme activity can convert nitriles to speciality or commodity chemicals, agrochemicals or pharmaceutical intermediates.[19] One of the most successful examples of biocatalytic production is the conversion of acrylonitrile to acrylamide using a biocatalyst obtained from *Rhodococcus rhodochrous* J1.[20] Polyacrylamide-immobilized J1 cells are used in a series of fixed-bed reactors to produce acrylamide from acrylonitrile at 10°C. In contrast to the traditional copper catalyst process, the bioprocess is carried out at room temperature and atmospheric pressure; it eliminates heavy metals and produces less process wastewater.

Lonza Guangzhou Fine Chemicals have developed a process to manufacture nicotinamide by chemoenzymatic route.[21] 2-Methyl-1,5-diaminopentane, a byproduct of the nylon-6,6 process, is converted to 3-methylpyridine, which is then ammoxidized to 3-cyanopyridine, followed by hydrolysis to nicotinamide by using immobilized *Rhodococcus rhodochorus* J1 cells.[22]

5-Cyanovaleramide (5-CVAM) is a starting material for the herbicide azafendin. It is produced from adiponitrile (ADN), by a highly polluting and cumbersome route using MnO_2 as a catalyst, giving only 25 % conversion and it requires a difficult solvent extraction process. Several bacteria containing nitrile hydratase are identified as catalysts for the hydration of ADN, which regioselectively produced 5-CVAM in high yield in a process by Dupont. *Pseudomonas chlororaphis* B23 was identified as the regioselective catalyst for commercial production of 5-CVAM.[8]

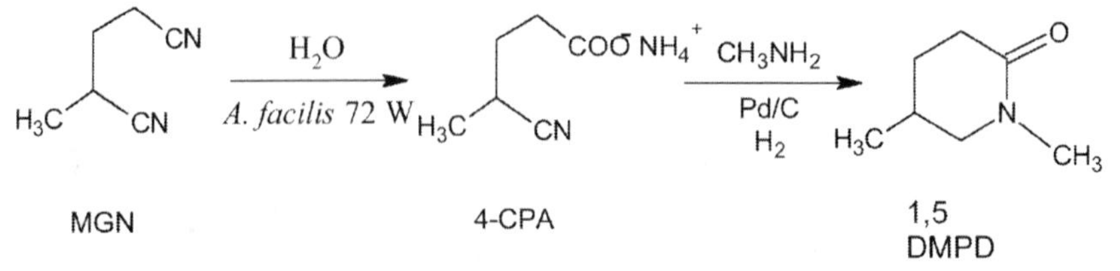

Figure 4. Chemoenzymatic conversion of 2-methylglutaronitrile (MGN) by *A. facilis* to form 1,5-dimethyl-2 piperidone (1,5-DMPD).

DuPont also claimed a commercial process for the conversion of 2-methylglutaronitrile (MGN, a byproduct produced during the manufacture of adiponitrile) to 1,5-dimethyl-2-piperidone (1,5-DMPD), which is used in electronics and coatings. *Acidovorax facilis* 72W cells containing nitrilase is used to hydrolyse MGN to 4-cyanopentanoic acid (4-CPA) ammonium salt (Fig. 4).[23] The reaction yields only one of two possible cyanocarboxylic acid ammonium salts with >98% selectivity at 100% conversion and produces only 1-2% of 2-methylglutaric acid diammonium salt as the only reaction byproduct.[24,25]

Aromatic Carboxylation

In the Kolbe Schimdt process, phenol on direct carboxylation gives p-hydroxybenzoic acid,[26] a compound used to produce liquid crystalline polymers. It produces dead end products and leads to the generation of solid waste of metal salts and tar residues. Phenol carboxylase has been used to produce 4-hydroxybenzoic acid in an enzymatic process.[27] A biological carboxylation route offers higher yields and lowers waste generation vis-à-vis the existing chemical method. A gene for the direct carboxylation has been identified.

Glucose to 1,3 Propanediol

The biological production of 1,3 propanediol (3G) from a renewable source such as glucose was claimed. However, it was not economical. Use of recombinant *E. coli* that contains two previously known metabolic pathways, one for the conversion of glucose to glycerol and another for that of glycerol to 3G is demonstrated. This biocatalyst can now be used to convert glucose to 3G in an economically attractive process.[28]

Pharmaceuticals and Drugs

Enzyme catalysis has a great impact on the production of single enantiomers of drugs of different classes.

Antihypertensive Agents

Antihypertensives are classified into various categories depending on their site of action and mechanism, such as β-adrenergic blocking agents (β-blockers), calcium channel blockers, angiotensin converting enzyme inhibitors, etc.

Atenolol, propanolol, metoprolol, penbuterol, etc. belong to the group of β-blockers. The β- blocking activity of these compounds resides mostly in (S)-enantiomer. Microorganisms are used as stereoselective epoxidation catalysts for the conversion of arylallylethers into (+) arylglycidylethers.[29] These ethers are then converted into S(−)-3-substituted-1-alkyl-amino-2-propanols, β-blockers. Chemoenzymatic synthesis of (S)-atenolol and (S)-propanolol also employs lipase catalyzed enantioselective esterification and hydrolysis reactions. Biocatalytic procedures for atenolol synthesis make use of lipase from *Pseudomonas cepacia, Candida cylindraceae* and *Mucor sp*. Among these *Pseudomonas cepacia* lipase has demonstrated an especially high stereoselectivity whereas *Candida cylindraceae* lipase shows opposite selectivity with low enantiomeric excess (ee) value.[30]

In the chemoenzymatic synthesis of the atenolol, the lipase catalyzes the deacylation of an intermediate, using 1-butanol in diisopropyl ether (DIPE) (Scheme1, Table 1).[31] The lipase from *Pseudomonas cepacia* shows excellent selectivity towards (S)—enantiomer giving complementary isomer (S) and (R) (*e.e.* = 100).

Captopril and enalapril are another group of antihypertensive drugs that inhibit the angiotensin-converting enzyme (ACE inhibitors) of the rennin-angiotensin system.[32-34] In the synthesis of captopril, the intermediate 3-acylthio-2-methyl propanoic acid ester is resolved to produce optically pure S-(−) enantiomer.[35,36] Lipases are used to catalyze the enantioselective hydrolysis of thioester bond of racemates.[37] Addition of L-proline to the (S)-enantiomer yields captopril (Scheme 2, Table 1).

Key chiral intermediates for the synthesis of monopril, a new antihypertensive drug, are synthesized chemoenzymatically. The asymmetric hydrolysis of 2-cyclohexyl-1,3-propanediol diacetate and 2-phenyl 1,3- propanediol diacetate to the corresponding (S)-2-cyclohexyl-1,3-propanediol

Table 1. Antihypertensives synthesized by chemoenzymatic routes

No.	Name of Drug	Scheme	Ref.
1	Atenolol	CH_2CO_2Bu … 1-butanol, DIPE / LIPASE → S (+) + R (-); R = H, = Ac	31
2	Captoptil	(R,S)-3-acylthio- 2-methyl propanoic acid ester —Lipase→ S(-)-3-acylthio- 2-methyl propanoic acid + R(+)-3-mercapto-2-methyl propanoic acid + Acetic acid	37
3	Monopril	2-cyclohexyl- 1,3-propanediol diacetate → (S)-2-cyclohexyl- 1,3-propanediol monoacetate; 2-phenyl 1,3- propanediol diacetate → (S)-2-phenyl- 1,3- propanediol monoacetate (Biphasic system, PPL or Candida viscosum lipase)	38
4	Enalapril	OAc, H, Ar, CN —Lipase→ OAc, H, Ar, CN + OH, H, Ar, CN	39
5	Diltiazem	OCH_3, CF_3 … —N. salmonicolor SC 6310→ … OH	40

monoacetate and (S)-2-phenyl-1,3-propanediol monoacetate is carried out by porcine pancreatic lipase (PPL) and *Chromobacterium viscosum* lipase (Scheme 3, Table 1).[38]

Ester hydrolase catalyzes resolution of cyanohydrin acetates to give corresponding α-hydroxy-carboxylic acids which can be used as valuable building blocks for numerous ACE inhibitors, e.g., Enalapril, as shown in Scheme 4, Table 1.[39]

Diltiazem, an antihypertensive, is a benzothiazepinone calcium channel blocking agent that inhibits influx of extracellular calcium through L-type voltage operated calcium channels. A key intermediate in the synthesis of diltiazem is obtained by an enantioselective microbial process. Enantioselective reduction of 4,5-dihydro-4-(4-methoxyphenyl)-6-(trifluoromethyl)-1H-11-benzazepin-2,3-dione, an achiral enol, gives only the single alcohol isomer (Scheme 5, Table 1). The most effective culture *Nocardia salmonicolor SC 6310*, catalyzes this conversion in 96% reaction yield with 99.8% ee.[40] There are two methods for the enzymatic resolution of the (R,S)-acid. One is the hydrolysis of chemically synthesized racemic ester and the other is direct esterification of racemic acid in organic media. Lipase is successful in resolving acids to produce S-acid in a single step.[41,42] Deracemisation of (±)-2-hydroxy-4-phenyl--butenoic acid, an important building block for ACE inhibitors, can be carried out by using a lipase. The nonreacting enantiomer is racemised with mandelate racemase. The outcome of this reaction is controlled by switching between the lipase catalysed acyl transfer or ester hydrolysis.[43]

Lipase from *Candida antartica* (fraction B) (CAL-B) is employed for the hydrolytic resolution of (R,S)-glycidyl butyrate at 25°C, pH 7 and 10% of dioxane to convert (R)-glycidyl butyrate (ee > 90% at 64% of conversion). In contrast to the enantiopreference displayed by most of the lipases, CAL-B is very specific towards hydrolysis of R-isomer. The reaction products can readily be converted to (R)-and (S)-glycidyl tosylates, which are very attractive intermediates for the preparation of optically active β-blockers and a wide range of other products.[44]

Non-Steroidal Anti-Inflammatory Drugs (NSAID)

2-Aryl-propionic acids are an important class of NSAIDs widely used for alleviating pain and inflammation associated with the tissue injury. Anti-inflammatory and analgesic effect is attributed almost exclusively to (S)-isomer.[45,46] Biocatalytic methods are developed to obtain ketoprofen, naproxen, ibuprofen, flurbiprofen in optically pure form.

(R,S) Flurbiprofen is resolved by using different lipases like *Candida rugosa*, *Mucor javanicus*, and porcine pancreas.[47] Lipase in presence of methanol causes esterification of (R,S)-Flurbiprofen, giving R-Flurbiprofen methyl ester and S-Flurbiprofen. (Scheme 1, Table 2).

Hydrolysis of racemic (R,S)-ketoprofen ethyl ester is carried out by using *Candida rugosa* lipase (CRL) to obtain (S)-ketoprofen. The enantioselective CRL enables the production of optically pure (S) ketoprofen[48] (Scheme 2, Table 2). A novel lipase from *Acinetobacter sp.* ES-1 shows a good enantioselectivity towards the resolution of (R,S)-ketoprofen ethyl ester to produce pharmacologically active (S)-ketoprofen. Aqueous phase reaction system gives ee of 99% at a conversion of 49% in 72 h.[49]

(S)-Ketoprofen is produced by a novel biphasic enzymatic membrane reactor (EMR). Enantioselective esterification and dynamic kinetic resolution are coupled under a single operating EMR. Figure 5 illustrates the process which includes esterification of (R,S)- ketoprofen acid in presence of lipase and alcohol to give (R)-ester. (R)-Ketoprofen ester is recovered in the organic phase while unreacted (S)-acid is separated in the aqueous layer. The second step in this process is the dynamic kinetic resolution i.e., combination reactions of in situ racemization of (R)-ester and hydrolysis to give (S)-acid.[50]

(S)-Naproxen is widely used for the treatment of disease such as arthritis and the (S)-form is 28 fold more active than the (R)-form. (S)-Naproxen is obtained from the racemic naproxen methyl ester by enantioselective hydrolysis by lipase (Scheme 3, Table 2).[51] The extracellular fungal lipases derived from genera *Rhizopus*, *Mucor* and *Candida* are uniquely enantiospecific in catalyzing the hydrolysis of methyl ester of naproxen. However, most lipases preferentially cleave the (R)-enantiomer; only the lipase from *Candida cylindracea* possess the desired (S)-stereochemical preference. Racemic naproxen amide can be enantioselectively hydrolyzed to form (S)-naproxen with resting cells of *Rhopdococcus erythropolis* (Scheme 4, Table 2).[52]

Table 2. Enantioselective synthesis of various anlagesics

No.	Name of Drug	Scheme	Ref.
1	Flurbiprofen	(R,S)-flurbiprofen → (*Candida antartica* lipase, Methanol, $-H_2O$) → R-flurbiprofen ester + S-flurbiprofen	45
2	Ketoprofen	(R,S)-Ketoprofen ethyl ester → (Lipase) → (S)-Ketoprofen + (R)-Ketoprofen ethyl ester	49
3	Naproxen	R,S - Naproxen ester → (Novozym 435, H_2O, 50°C, -ROH) → R-Naproxen acid + S-Naproxen ester	51
4	Naproxen	R,S - Naproxen ester → (Amidase) → R-Naproxen amide + S-Naproxen	52

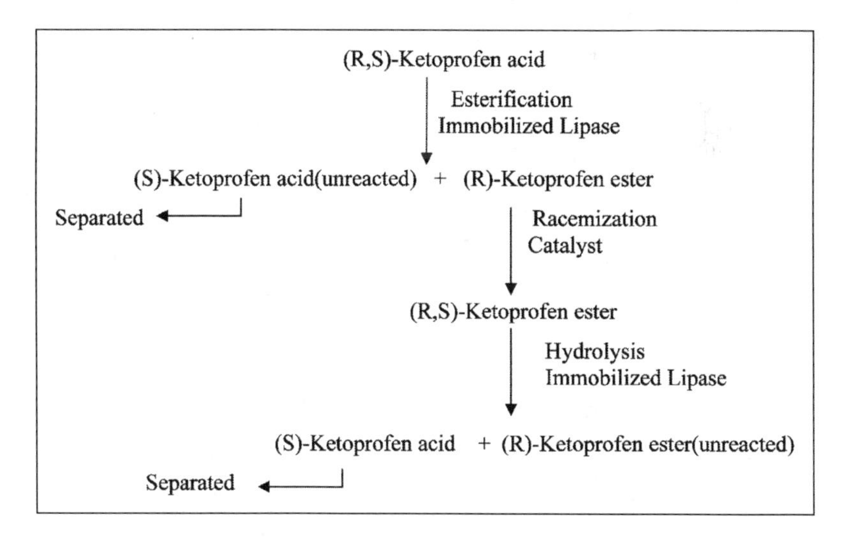

Figure 5. Schematic representation of biphasic EMR in production of (S)-Ketoprofen.

Antimicrobial/Antibiotic Agents

Antimicrobials or antibiotics are the substances produced by microorganisms which have the capacity of inhibiting the growth and even of annihilating other microorganisms. Cephalosporins are mainly produced by chemical acylation of acyl acceptor with acyl donors.[53,54] The multiple step chemical synthesis of cephalosporins involve side chain protection, carboxyl group activation, condensation reaction and side chain deprotection carried out in methylene chloride. Because of the traces of the carcinogenic contaminants in antibiotics produced by chemical methods and environmental factor, there is a need to develop cleaner methods. The potential alternative is the enzymatic synthesis catalysed by penicillin G acylase in aqueous medium under mild conditions. Enzymatic synthesis of cephalexin has also been reported.[55-57] Penicillin G acylase is a water soluble enzyme and it prefers catalyzing reaction in aqueous environment. The higher solubility of 2-thienyl acetic acid and 2-thienylacetohydroxamic acid (2-TAH) in aqueous solution makes them better substrates for penicillin G acylase. Scheme 1 in Table 3 represents the enzyme mediated step in the synthesis of the cephalothin by penicillin G acylase (PGA).[58] Kinetic studies of the synthesis of β-lactam antibiotics by using PGA has suggested that PGA may act either as a transferase or as a hydrolase, catalyzing two undesired side reactions, namely, the hydrolysis of the acyl side-chain precursor (an ester or amide, a parallel reaction) and the hydrolysis of the antibiotic itself (a consecutive reaction).[59]

The first broad spectrum antibiotic, chloramphenicol, initially isolated from cultures of various streptomyces strains was also the first to be produced industrially by chemical synthesis.[60,61] Now it is produced chemoenzymatically as shown in Scheme 2, Table 3.[62] The process involves enantioselective hydrolysis of methyl (±) threo-N-dichloroacetyl-β-(4-nitrophenyl) serinate by subtilisin. This enzymatic resolution gives the corresponding (2S,3R)-acid and the unhydrolysed

Table 3. Chemoenzymatic synthesis of antimicrobials

Sr. No.	Name of Drug	Scheme	Ref.
1	Cephalosporin	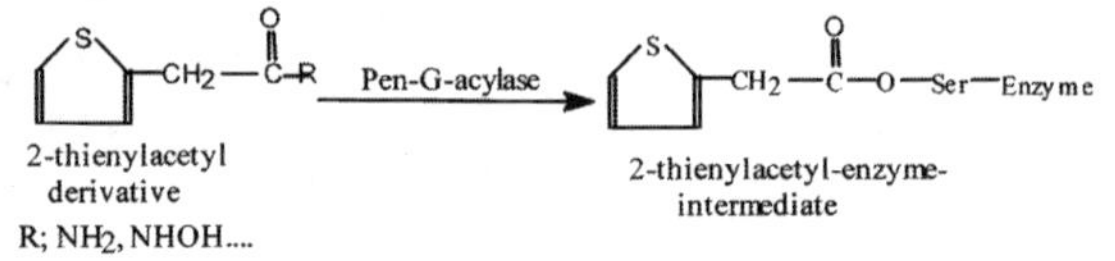 2-thienylacetyl derivative, R; NH_2, NHOH.... —Pen-G-acylase→ 2-thienylacetyl-enzyme-intermediate (CH_2—C(=O)—O—Ser—Enzyme)	58
2	Chloramphenicol	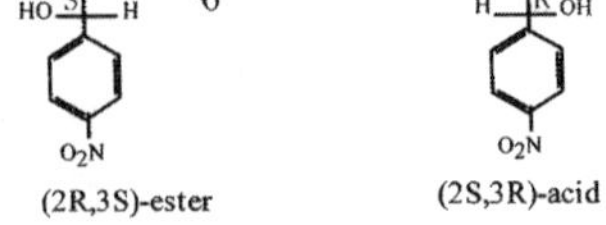 Subtilisin → (2R,3S)-ester + (2S,3R)-acid	62
3	Cephalopsorin	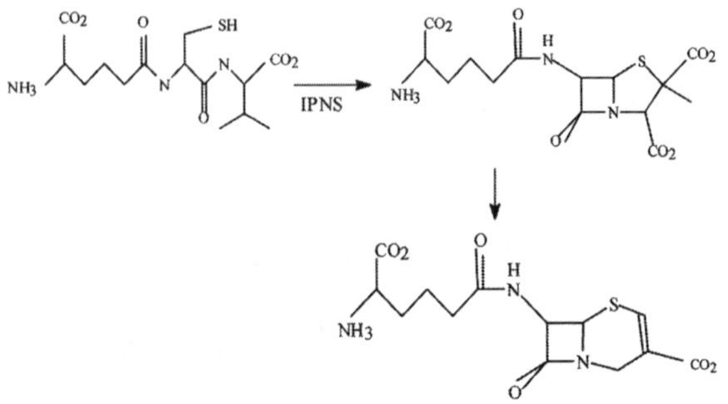 —IPNS→	63

(2R,3S)-ester(−)-threo in high yield and high optical purity. Reduction of the acid by borane-methylsulfide complex gives chloramphenicol(+) and that of the ester by lithium aluminium hydride gives the chloramphenicol (−).

The tripeptide aminoadipoyl-cysteinyl-D-valine (ACV) is oxidatively transformed to the 4-5 bicyclic β-lactam ring system by isopenicillin N synthase (IPNS). IPNS is a member of substantial family of iron-containing enzymes that use Fe^{2+} to activate both O_2 and specific cosubstrate for complex redox chemistry (Scheme 3, Table 3).[63-65] The five membered ring in penicillins can be expanded to the six membered ring in cephalosporin antibiotics. Several natural products contain five-membered-ring heterocycles (oxazoles and thiazoles) that arise from enzymatic cyclization of serine or cysteine residues in peptide precursors.[66] These include the *E. coli* antibiotic microcin B17, which has the same mode of action as ciprofloxacin.[67]

Antidepressant and Antipsychotic Agents

Tomoxetine, fluoxetine and paroxetine are some of the popular antidepressants. The (R)-isomer of tomoxetine is nine times more potent than the (S)-isomer. Fluoxetine is active against a wide range of symptoms like anxiety, alcoholism, chronic pain, obesity and bulimia.[68] Synthesis of (R)-Fluoxetine has been widely attempted. Enantioselective epoxidations, reductions, hydroborations and others are described in literature for the preparation of suitable chiral building blocks for these materials. Enzymatic synthesis of enantiopure benzyl alcohols is extended to develop a practically useful and highly efficient method for the lipase catalyzed resolution of (R) and (S)-3-chloro-1-phenyl-1-propanols which are used in the synthesis of antidepressants by a combination of direct substitution and "Mitsonubu" inversion (Scheme 1, Table 4).[69]

Chemoenzymatic enantioselective synthesis of tomexetine and fluoxetine is also possible from commercially available inexpensive starting material ethyl benzoyl acetate. Baker's yeast reduction of ethyl benzoyl acetate to optically pure (−)-N-methyl-3-phenyl-3-hydroxy propylamine (Scheme 2, Table 4) serves as a key step in the synthesis of R(−) tomoxetine, (R)- and (S)-fluoxetine.[70]

Table 4. Two different chemoenzymatic route for synthesis of fluoxetine and buspirone

Sr. No.	Name of Drug	Scheme	Ref.
1	Fluoxetine	OCOCH₂Cl … Cl → OH … Cl + OCOCH₂Cl … Cl	69
2	Fluoxetine	Ethyl benzoyl acetate —Baker's yeast reduction→ 3-hydroxy-3-phenyl-propionate	70
3	Buspirone	(R,S) 6-acetoxy buspirone —Amano acylase 30000→ (S)-6-Hydroxy buspirone + (R)- 6-acetoxy buspirone + acetate	71
4	Buspirone	—Streptomyces antibioticus ATCC 14890→	72

6-Hydroxy buspirone, an active metabolite, is synthesized in enantioselective way by three different routes, namely, (i) resolution of racemic 6-acetoxybuspirone to (S)-6-hydroxybuspirone using L-aminoacid acylase from *Aspirgillus melleus* (Amano acylase 30000), at 45% conversion with 95% ee (ii) *Streptomyces antibiotics* ATCC 14890 catalysed hydroxylation of buspirone, and (iii) microbial reduction of 6-oxo-8-[4-[4-(2-pyrimidinyl)-1-piperazinyl]butyl]-8-azaspiro[4,5]-dcane-7,9-dione to (R)- or (S)-6-hydroxy-8-[4-[4-(2-pyrimidinyl)-1-piperazinyl]butyl]-8-azaspiro[4,5]-decane-7,9-dione.[71,72]

Antitubercular Agents

(+)-2-Amino-1-butanol is an important intermediate for producing (+)-2,2ǀ-(ethylenediimino)-di-1-butanol i.e., ethambutol, which has a remarkable antitubercular action. The racemic form or meso form 2,2ǀ-(ethylenediimino)-di-1-butanol produces strong undesirable side effects on the eyes. Thus an extremely high optical purity is required for (+)-2-amino-1-butanol. The chemical process involved in the production of (+)-2-amino-1-butanol, from 1-nitropropane gives a yield as low as 10%. Furthermore, the simultaneously produced (−)-2-amino-1-butanol is quite difficult to be racemized making it unviable. A newer enzymatic method was used for the production of this intermediate through the resolution of N-acyl-DL-2-aminobutyric acid in aqueous phase by using an amino acylase. (+)-2-Amino-1-butanol is then obtained by esterifying L-2-aminobutyric acid which is obtained by the optical resolution of DL-2-aminobutyric acid and reducing the resulting L-2-aminobutyric acid ester (Scheme 1, Table 5).[73] The reaction of ethyl isonicotinate (ethyl 4-pyridine carboxylate) with hydrazine hydrate as a nucleophile has been studied in an organic solvent with an immobilized lipase Novozym 435 as catalyst to produce 4-pyridine carboxylic acid hydrazide (Isoniazid) which is an important agent in the treatment of tuberculosis.[74]

Anticancer Agents

Microtubule proteins are responsible for the formation of the spindle during cell division. Anticancer agent that act on these proteins are grouped under the class 'spindle formation inhibitors', such as vinblastin, colchicines and paclitaxel.[75,76] Paclitaxel (taxol), a complex polycyclic diterpene is the only compound which exhibits unique mode of action on these proteins.[77,78] Originally paclitaxel was produced from extracts of the bark of the Pacific yew tree by a cumbersome purification process in relatively low yields. Alternative methods include cell suspension cultures and semisynthetic chemoenzymatic synthesis.[79-82] The latter method involves the formation of side chain synthon enzymatically. The stereoselective hydrolysis of racemic acetate (*cis*-3-acetyloxy-4-phenyl-2-azetidinone) to yield the desired (3R-*cis)*-acetyloxy-4-phenyl-2-azetidinone and (3S-*cis*)-hydroxy-4-phenyl-2-azetidinone by lipase has been demonstrated (Scheme 1, Table 6).[83]

Table 5. Chemoenzymatic synthesis of antitubercular drug

No.	Name of Drug	Scheme	Ref.
1	Ethambutol	COOH, NHAc (±)-2-Aminobutric acid acetate —Amino acylase→ COOH, NH_2 (+)-2-Aminobutric acid + COOH, NHAc (-)-2-Aminobutric acid acetate	73
2	Isoniazid	O ‖ C—OC_2H_5 (N) + NH_2NH_2 —Lipase→ O ‖ C—$NHNH_2$ (N) + C_2H_5OH	74

Table 6. Chemoenzymatic synthesis of anticancer drug

Sr. No.	Name of Drug	Scheme	Ref.
1	Anticancer	OAc, NH — Lipase → OAc, NH + OAc, NH Racemic acetate 3R-Acetate 3S-Alcohol	83

Table 7. Chemoenzymatic synthesis of antifungal drug

Sr. No.	Name of Drug	Scheme	Ref.
1	Antifungal	F, F, O—C(=O)—OR, OH, O—C(=O)—OR — Lipase → F, F, O—C(=O)—OR, OH, OH Prochiral diester Optically active diol	84

Antifungal Agents

Two new antifungal agents, ZD0870 and Sch45450 show broad spectrum antifungal activities compared to other clinically used antifungal agents. Higher activity is attributed to chiral centers comprised of a difluorobenzene and a triazole group. An optically active diol is a useful common intermediate for their synthesis. Antifungal agents, chiral azoles, can be prepared by chemoenzymatic process only in five steps. The clinically active diol can be obtained by using an enzymatic stereoselective hydrolysis of prochiral diesters as a key reaction (Scheme 1, Table 7).[84]

Antiulcerant Agents

Antiulcer agents are the compounds which act as active H^+K^+ATPase inhibitors. They are effective acid secretion inhibitors. These compounds include omeprazole, lansoprazole and pantoprazole, etc. Antiulcer agents being sulfoxides, have an asymmetric centre in the sulfur atom. Inorganic and organic sulfur exhibit a wide range of stable oxidation states which are readily interconvertible, making these compounds useful in oxidations and reductions.[85] Single enantiomer of sulfoxide compounds can be obtained by biooxidation of the prochiral sulfide moiety. Enzymatic sulfur oxygenation reactions are of two types, involving dioxygenases and monooxygenases.

Synthesis of enantiomerically pure (S)-isomer of omeprazole involves the oxidation of the sulphur of the prochiral intermediate using the monooxygenase obtained from different microbial sources (Scheme 1, Table 8).[86] Oxidase/peroxidase coupled enzyme systems can also be used for the oxidation of the heteroaryl-methyl-sulfides to produce corresponding chiral sulfoxides.[87]

Miscellaneous

Isosorbide-5-nitrate and isosorbide-2,5-dinitrate are used for the treatment of angina pectoris. The synthesis of these molecules involves selective functionalization of isosorbide prior to the introduction of the nitro group. It can be achieved with high efficiency and complementary stereochemistry by using suitable conditions with the highly selective lipase from *Pseudomonas sp.*[69] 2,5-Diacyl derivative of isosorbide undergoes selective hydrolysis at 5-C to produce 2-acyl

Table 8. Chemoenzymatic synthesis of Omeprazole

Sr. No.	Name of Drug	Scheme	Ref.
1	Omeprazole	H_3C, OCH_3, CH_3, N, S, N, NH, OCH_3 → monooxygenase → H_3C, OCH_3, CH_3, N, S, O, N, NH, OCH_3; S-Omeprazole	86

derivative. Simple nitration followed by removal of the acyl group leads to the desired enantiomerically pure pharmaceuticals (Scheme 1, Table 9).

Indinavir is an antiviral drug (anti HIV drug). Synthesis of its intermediates is achieved by a chemoenzymatic process. The synthesis of 1*S*,2*R*-1-amino-2-indanol, which is a key component of HIV protease inhibitor, is achieved in four steps starting from indanone efficiently and with high levels of diastereo- and enantioselectivity. The 2-acetoxy-1-indanone is hydrolyzed to 2-hydroxy-1-indanone enantioselectively by using *Rhizopus oryzae* (Scheme 2, Table 9).[88] 2-Hydroxyl-1-indanone is an intermediate in the synthesis of Indinavir.

Lipases can also be used in the resolution of secondary alcohols. Thus (±) 3,2-Cl (NC)-$C_6H_3OCH_2CH(OH)CH_2Cl$ is resolved by using lipase catalysed transesterification with vinyl acetate to produce calcylitic agent NPS-2143 and it is used in the osteoporosis treatment.[89]

Lipase PS-30 from *Pseudomonas cepacia* can be used in the preparation of a chiral synthon for an Hepatitis B Virus (HBV) inhibitor. Enantioselective enzymatic asymmetric hydrolysis of (1α,2β,3α)-2-(benzyloxymethyl)cyclopent-4-ene-1,3-diol diacetate to the corresponding (+)-monoacetate gives a yield of 85% with an ee of 98% (Scheme 3, Table 9). Pancreatin gives a yield of 75% with an ee of 98.5%.[90]

Table 9. Schemes of the drugs synthesized chemoenzymatically

Sr. No.	Name of Drug	Scheme	Ref.
1	Vasodialator	OAc, H, 5, O, O, 2, H, OAc → Lipase from *Pseudomonas fluorescence*, Buffer pH 7, THF → OH, H, 5, O, O, 2, H, OAc; **2,5-diacyl derivative of isosorbide** → **2-acyl derivative**	69
2	Indinavir	O, OAc → Lipase from *Rhizopus oryzae* → O, OH + O, OAc; **2-acetoxy-1-indanone** → **2-hydroxy-1-indanone** + **2-acetoxy-1-indanone**	89
3	HBV inhibitor	OBn, AcO, OAc → Lipase PS-30 from *Pseudomonas cepacia* or Pancreatin → OBn, HO, OAc; **Diacetate** → **(+)-Monoacetate**	90

Strategies of Enhancing Conversions of Chiral Compounds

In the case of chiral compounds that contain equal proportion of (R) and (S) isomer, enzymes catalyze the transformation of only one of the isomers preferentially leading to a maximum of only 50% conversion. Thus, these kinetically controlled processes cause 50% loss of the valuable reactant. Actual yields obtained in most of the cases are very low because of further losses during separation processes. Thus, transformation of a racemate into a single stereoisomer, the deracemization process, is highly desirable and typically kinetic resolution is employed to increase yield in this process. The classical kinetic resolution requires some additional processes for the recycling of unwanted enantiomers: separation, racemization and repeated resolution. One approach to overcome this limitation, and thus to increase the yield up to 100%, is to employ a racemization catalyst for the in-situ conversion of less reactive enantiomers to more reactive enantiomers together with the enzyme. So to make the kinetic resolution economic, it should be possible to use the nonreacted isomer for some other application. This limitation of the enzymatic process can be overcome with dynamic kinetic resolution (DKR) and enantioselective enzymatic desymmetrization (EED). The EED implies the modification of symmetric compounds that eliminates one or more elements of symmetry of the substrate and DKR consists of carrying out an in-situ continuous racemization of the substrate so that theoretically all of the starting racemic material can be used for transformation into one enantiomer. As a case with chemical methods of EED, enantioselectivity can be achieved using enzymes if the symmetry elements that preclude chirality are eliminated.[91]

Enantioselective Enzymatic Desymmetrization (EED)

EED, a very interesting alternative to kinetic resolutions (KR) belongs to the field of synthesis and accordingly a maximum yield of 100% can be obtained.[92] It involves the desymmetrisation of meso and prochiral compounds. Some of the recent examples of EED are summarized in Table 10. Not only polyhydroxylated straight chain compounds have been desymmertrized but also the meso diols including three membered ring,[93] 2,6 and 2,4,6 substituted piperidines,[94] polyhydroxylated cyclohexanes,[95] and triols[96] using EED in high enantiselectivity and better yield. A challenging issue in the asymmetric synthesis is the transformation of the polyhydroxylated cyclohexanes which can be efficiently done using enzyme approach.[97] Other than lipases and esterases, nitrile hydratases and amidases are studied for the EED of dinitriles in which one of the nitrile groups converted to acid through the amide formation.[98-100] Nitrile hydratase converts nitrile to amide by hydration and the amide is then transformed to the acid because of the action of amidase.

Baker's yeast (*Saccharomyces cerevisiae*) has been the most popular whole-cell biocatalyst for the desymmetrization of prochiral ketones[101] which follows Prelog's rule. Although a lot of efforts have been expended in the optimization of microorganisms that catalyze reduction of different ketones to obtain both enantiomers of secondary alcohols, EED of small dialkyl ketones still remains a major challenge. Towards this goal a dimorphic fungus *Geotrichum candidum* IFO 4597 works well.[102,103] With this species better yields were obtained with *ortho*-substituted acetophenones than *para*-substituted acetophenones, for the oxidation of the product formed is not possible for *ortho*-substituted compounds. Large scale application using alcohol dehydrogenases suffers from the drawback of the instability of the biocatalyst since it requires high concentrations of the cosubstrate like 2-propanol to drive the reaction to completion. Microorganism *Rhodococcus ruber* is an exception which is stable towards organic solvents (22% v/v 2-propanol) and catalyse the reduction of broad range of ketones.[104] One of the approaches to increase the yield and ee of these reactions is to employ a surfactant like sodium lauryl sulfate under anaerobic condition. In this way bromoalcohols have been produced and used in the synthesis of salmeterol.[105,106] Not only microbes but also the whole plant cells can be used for the purpose of reduction reactions.[107] In most of the cases, whole cells are used as catalyst for oxidation reduction reactions, which poses problems due to the presence of other enzymes. Inhibition of a particular enzyme is the solution to it as described by Yadav et al.[108] Some other organisms that are employed for reduction reactions are *Zygosaccharomyces bailii,*[109] *Torulaspora delbrueckii,*[110] thermophyllic actinomycete, *Streptomyces thermocyaneviolaceus,*[111] *Aureobasidium pullulans,*[112] ADH from *Lactobacillus brevis* expressed in *E. coli.*[113,114]

Table 10. Desymmetrization of different compounds using enzymatic methods

Catalyst	Reaction Condition	Reaction Type	Substrate	Yield (%)	ee (%)	Comments	Ref.
Pseudomonas fluorescence Lipase	RT, 4 h	Transesterification (vinyl acetate)	2-ethyl-1,3-propanidiol	72 (R)	46	Enzyme enantioselectivity higher for hydrolytic transformation	117
Pseudomonas fluorescence Lipase	RT, 1.5 h	Hydrolysis	di-O-acetate of 2-ethyl-1,3-propanidiol	65 (R)	94		
Porcine Pancreas Lipase	30°C, 2 h	Transesterification (vinyl acetate)	N-BOC protected serinol	69 (R)	>99	Self correcting process. Depending upon the reaction condition the diacetate formed from the unwanted enantiomer of the monoacetylated product	118
Pseudomonas Sp. Lipase	25°C	Transesterification	2-carbamoylmethyl-1,3-propanediols (N-alkyl derivative)	26 (S)	>99	1. Low yield due to diacetal formation. 2. Change in enantiselectivity based upon substitution	119
			2-carbamoylmethyl-1,3-propanediols (N,N-dialkyl derivative)	77 (R)	96		

continued on next page

Table 10. Continued

Catalyst	Reaction Condition	Reaction Type	Substrate	Yield (%)	ee (%)	Comments	Ref.
Candida antartica Lipase B	2 h	Transesterification	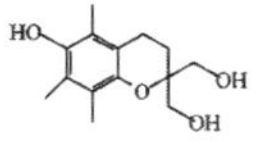 Bicyclic triol	1. Mono -ester, 60% (S). 2. Diester, 27%	98 –	1. Intermediate in S-α-tocotrienol. 2. Yield can be increased by the non enzymatic hydrolysis of diester	120
Pseudomonas Sp. Lipase	RT, 9.5 h, Phosphate buffer pH 7	Hydrolysis	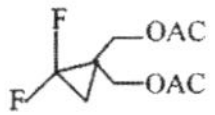	68 (S)	93.5	Changing selectivity on the basis of reaction type	121
	35°C, 1.5 h	Transesterification	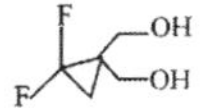	94 (R)	91		
Candida rugosa Lipase	RT, 4 h	Transesterification	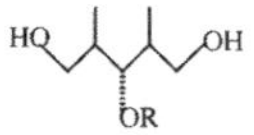	94 (2R, 3R,4S)	97	Intact sieves to trap byproduct acetaldehyde to achieve high ee	122
Porcine Pancreas Lipase	RT, 24 h	Transesterification	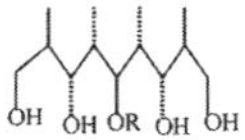	76 (2R, 3R,4S,5R, 6R,7S,8S)	>98	1. C19-C21 fragment of rifamycin S. 2. Enzyme highly regioselective for a primary alcohol end group	123
α-chymotrypsin	pH7, phosphate buffer, 24-30 h	Hydrolysis	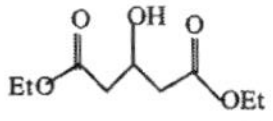	96 (S)	98.1	Building block of atorvastatin	124
Rhodococcus sp CGMCC 0497	Phosphate buffer, pH 7, 30°C, 24h	Hydration	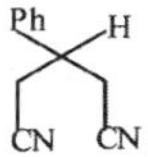	53 (S)	>99	Acetone used as additive interacts with the biocatalyst and enhances the enantioselectivity	125

continued on next page

Table 10. Continued

Catalyst	Reaction Condition	Reaction Type	Substrate	Yield (%)	ee (%)	Comments	Ref.
Geotrichum candidum	30°C, 24h, XAD-7	Reduction		74-92 (S)	92->99	1. Yield and ee depends upon the R group 2. Enantiselectivity depend upon the reaction condition	126
	30°C, 24h, Aerobic condition			56-99 (R)	85->99		
Rhodotorula rubra	SLS, 30°C, 48h argon	Reduction		79 (R)	95	Addition of SLS and anaerobic condition increases the yield and ee	
Rhodotorula mucillaginosa	28°C, 48h	Reduction		88 (R)	>99	Different organisms used to obtain the different isomer from the same substrate	127
Geotrichum candidum	28°C, 48h XAD-1180			95 (S)	>98		
Geotrichum sp.	30°C, 15h	Reduction		77 (R)	60	Enantioselectivity and yield depends upon the position of the substitution	128
	30°C, 12h			91 (S)	96		
Acetobacter pasterurianus DSM 8937	Glycerol, pH 7.2, 32°C, 2h	Oxidation		100 (R)	97		115

Enantioselective oxidation of meso and prochiral diols to produce chiral compounds is the reverse way of utilizing these biocatalysts. For example, (R)-3-hydroxy-2-methyl propionic acid, a key intermediate of the synthesis of captopril has been prepared by oxidative EED of prochiral 2-methyl-1,3-propanidiol.[115] Biocatalytic methods are not limited to the desymmetrization of compounds containing hydroxyl, keto, cyanide groups but applied to the prochiral alkenes as well. Organisms which have this ability are *Rhizopus arrhizus, P. fluorescens* and *S. cerevisiae*, of which *S. cerevisiae* have been studied widely.[116]

Dynamic Kinetic Resolution (DKR)

The kinetic resolution (KR) combined with in-situ racemization is called dynamic kinetic resolution (DKR).[129-133] For the DKR reaction to approach 100% conversion to a 100% enantiomerically enriched product rate of racemization reaction should be higher than that of the enzymatic resolution reaction. A racemization catalyst may be other enzyme (racemase), a base or metal complex as summarized in Table 11.

The most common strategy is perhaps to combine the enzyme with a nonmetallic base catalysed racemization. However, this strategy is mainly limited to substrates possessing a stereogenic centre with an acidic proton. Production of (S)-suprofen in 95% ee from (R,S)-suprofen 2,2,2-trifluoroethyl thioester in isooctane by using lipase catalysis and in-situ racemization with trioctylamine as the catalyst was developed by Lin et al.[140]

The consideration of compatibility between the enzymatic KR and the racemization process is very important. However, other equally important parameters are solvent, metal catalyst, nature of acyl donor, and temperature.[144,145] Racemization catalysts like ruthenium are not compatible with conventional acyl donors because acetaldehyde or acetone, which is formed in the acylation step, oxidizes the substrate. The problem was solved by employing *p*-chlorophenyl acetate (PCPA) as the acyl donor. In the presence of ruthenium catalysts as racemizing agents dynamic lipase resolution have furnished products in high yield and >99% ee.[146] But the unreacted PCPA makes the separation of acylated products problematic during work-up. Aromatic alcohols are better substrates for DKRs because they undergo rapid racemization with metal catalyst. One interesting process of the DKR is the coupling of the metal catalysed reduction of the ketone and the resolution/racemization of the alcohol thus produced with lipase-metal complex to avoid the use of cofactor-requiring enzymes. Similarly enol acetates are converted to single enantiomers involving four steps, the first being the lipase catalysed hydrolysis to produce ketone.[147] The racemization is proposed to take place within the coordination sphere of the ruthenium catalyst.[139] A key intermediate, (S)-5-(*tert*-butyldimethy

Table 11. Combinations of racemization and resolution catalyst with there substrate specificity

Sr. No.	Combination	Substrate	Ref.
1.	Lipase-palladium	Allyl acetates, amines	134,135
2.	Lipase-rhodium	Secondary alcohols	136
3.	Lipase-ruthenium	Secondary alcohols, ketones and their enol esters	137
	1. Indenyl-ruthenium		
	2. Pentaphenylcyclopentadienyl ruthenium		134 138
	3. (*p*-cymene)-ruthenium		139
4.	Lipase-Trioctylamine	Thioester	140
5.	Lipase-Acid zeolites like H-Beta	Secondary alcohols (benzylic alcohols)	141
6.	Racemase with other enzymes	Aminoacids	142
7.	Lipase-basic resin (amberlite OH^-)	Cyanohydrins	143

lsiloxy)heptanal, in the synthesis of widely used commercial insecticide Spinosyn A produced with ee up to 99% and conversion up to 92% using ruthenium-lipase catalyzed DKR.[148] Gutierrez et al have described the DKR process for an enantioconvergent microbial Baeyer-Villiger oxidation of benzyloxycyclo pentanone, to make use of whole cell to carry out kinetic resolution and an anion exchange resin to catalyze in-situ racemization.[149]

Biocatalysis in Nontraditional Media

Biocatalysis in non-aqueous environments has been widely discussed for more than two decades and its importance and applicability well-recognized. Ionic liquids and supercritical fluids have emerged as new attractive non-aqueous reaction media for biocatalysis.

Ionic Liquids

Ionic liquids are generally recognized as green solvents, have unique properties such as non-volatility, nonflammability and excellent chemical and thermal stability. They have low melting point (<100°C) and remain as liquids within a broad temperature range, and possess negligible vapour pressures. They represent an ideal reaction medium for chemical and biochemical reactions because of their highly polar nature. They are able to dissolve a wide range of different substances including polar and nonpolar organic, inorganic and polymeric compounds.[150]

Despite of their high polarity, most of ionic liquids are hydrophobic and can dissolve upto 1% of water and the presence of water may affect their physical properties. Enzymes normally do not dissolve in ionic liquids, but remain suspended as a powder or immobilized by a solid support. Varieties of enzymes are capable of performing catalytic activities (Table 12), which are generally comparable with or higher than those observed in conventional organic solvents. In ionic liquids, some enzymes are even more enantioselective and regioselective.[151]

Erbeldinger and coworkers reported the feasibility of using isolated enzymes in ionic liquids. Thermolysin catalyze synthesis of Z-Aspartame, which is an artificial sweetner. Scheme 1 in Table 13 gives synthesis of intermediate of Aspartame.[161]

Ionic liquids support a large and diverse range of organic reactions including oxidations, coupling reactions, nucleophilic displacements, reductions, and alkylations. Purified enzyme reactions are reported in ionic liquids. It is possible to conduct whole-cell biotransformations such as yeast mediated reductions of ketones in an ionic liquid. This combines advantages of whole cell bioreagents with the advantages of ionic liquids, principally their recyclable nature (refer to Table 14).[166]

Baker's yeast has been used to carry out a variety of transformations in synthetic organic chemistry. However, its use in this area has been limited by the necessity to employ aqueous solvent systems. Wide applications of these biocatalysts in synthetic chemistry can be obtained by retaining their selectivity in solvents more compatible with organic compounds. Examples of microorganism mediated reduction of ketones in ionic liquid is summarized in Table 14.

Table 12. Examples of enzyme catalysis in ionic liquids

No.	Biocatalyst	Reaction	Ref.
1	Lipase	Transesterification	152
		Kinetic resolution of chiral alcohols	153
		Resolution of amino acid ester	154
		Esterification of carbohydrates	155
2	Alcohol dehydrogenase	Enantioselective reduction of 2-octanone	156
3	Thermolysin	Synthesis of Z-Aspartame	157
4	Peroxidase	Oxidation of Guaiacol	158
5	Baker's Yeast	Enantioselective reduction of ketones	159
6	*Lactobacillus kefir* cells	Asymmetric reduction of 4-chloroacetophenone to (R)-1-(4-chlorophenyl)ethanol	160

Table 13. Lipase catalysis in ionic liquids

No.	Reaction	Ref.
1	L-Phenylalanine + Z-L-Aspartic acid → (Thermolysin, [bmim][PF_6]/H_2O 95:5 v/v, 37°C, 50h) → 95 % yield Aspartame precursor	162
2	Phenyl ethanol → (Lipase/24°C/3days; vinyl acetate (OAc) → CH_3CHO) → phenyl ethanol (OH) + Acylated product (OAc)	163
3	Cyclohexene → (Ionic Liquid, Yield 83%) → cyclohexene oxide; RCOOH → RCOH; H_2O_2 → H_2O (CaLB)	164
4	Octanoic Acid + NH_3 → (Cal B, Ionic liquid ([bmim][BF_4]) → octanamide (NH_2)	165

Supercritical Fluids

Recently supercritical fluids are employed as solvents in biocatalysis reactions. Among many fluids, supercritical CO_2, has the added benefits of an environmentally benign nature, nonflammability, low toxicity, high availability and an ambient critical temperature (T_C = 31°C).[169] Most of the enzymes used in supercritical fluids are hydrolytic enzymes such as lipases and proteases.[172,173]

***Table 14.* Different ketones, which undergo reduction in presence of ionic liquids**

Sr. No.	Ketone Substrate	Alcohol Product	Ref.
1	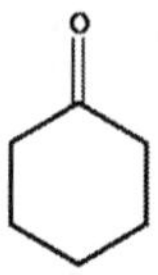	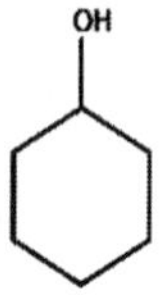	166
2	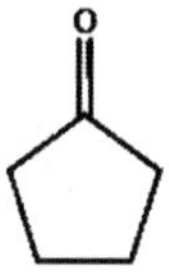	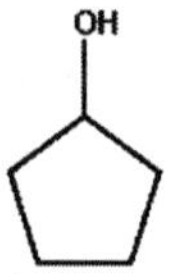	166
3	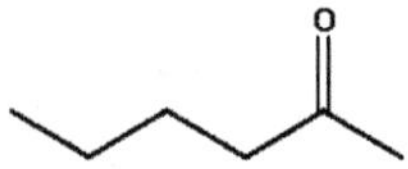	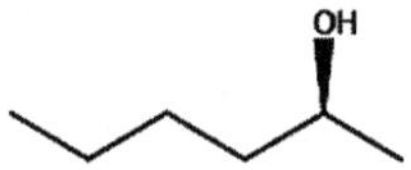	167
4	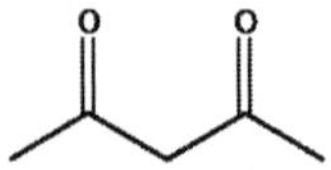	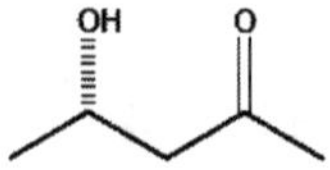	168
5	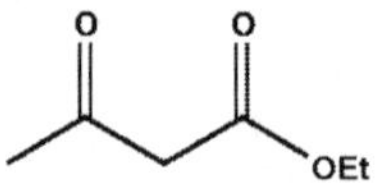	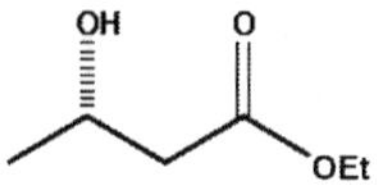	169
6	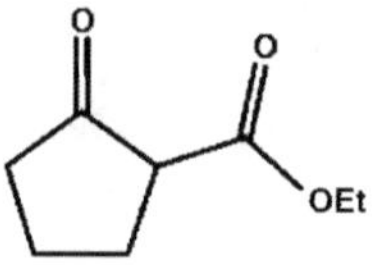	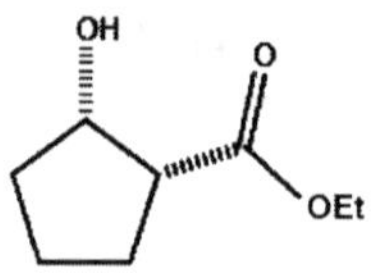	166
7	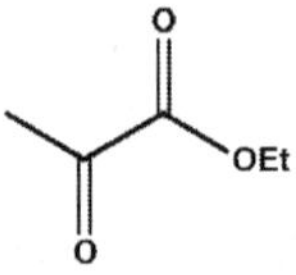	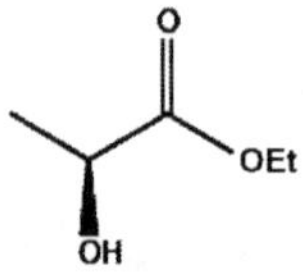	170

Enantioselective reduction of ketones in supercritical fluids is possible (see Table 15). The resting cells of a fungus, *Geotrichum candidum* were employed for the reduction of various ketones in supercritical CO_2 (Fig. 6).[178-180] Enzyme catalyzed reactions reported in supercritical CO_2 are given in Table 15.

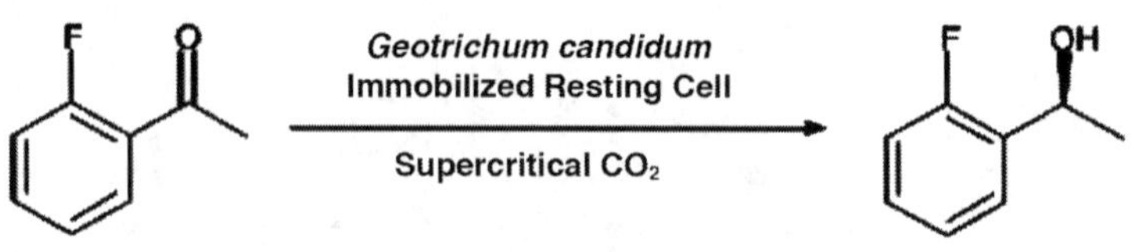

Geotrichum candidum
Immobilized Resting Cell
Supercritical CO_2

Figure 6. Whole cell catalyzed enantioselective reduction in supercritical fluid.

Table 15. Enzymatic reactions in supercritical CO_2

No.	Reaction	Ref.
1	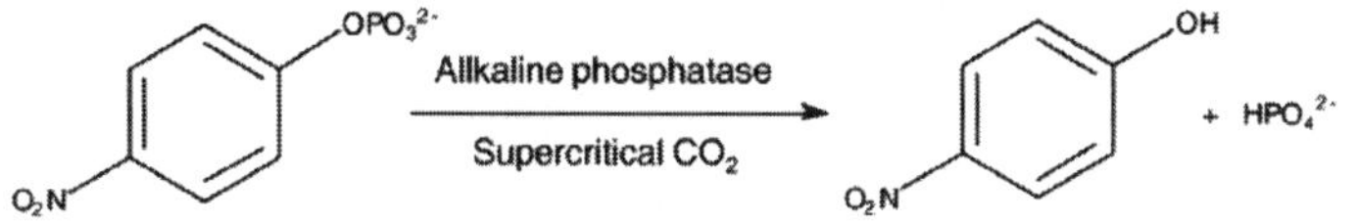 Alkaline phosphatase, Supercritical CO_2 (O_2N–C_6H_4–OPO_3^{2-} → O_2N–C_6H_4–OH + HPO_4^{2-})	174
2	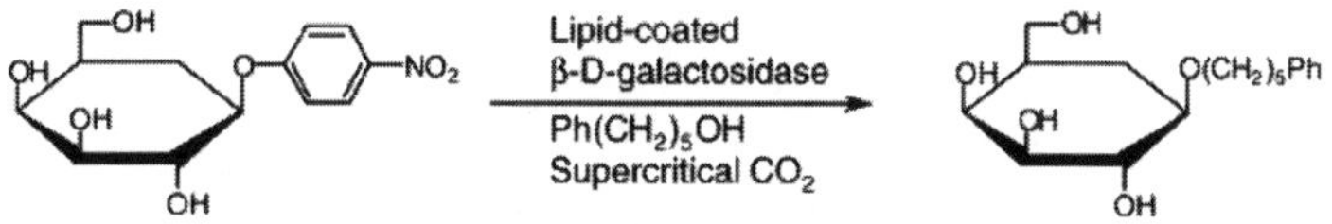 Lipid-coated β-D-galactosidase, $Ph(CH_2)_5OH$, Supercritical CO_2 (product: O$(CH_2)_5$Ph)	175
3	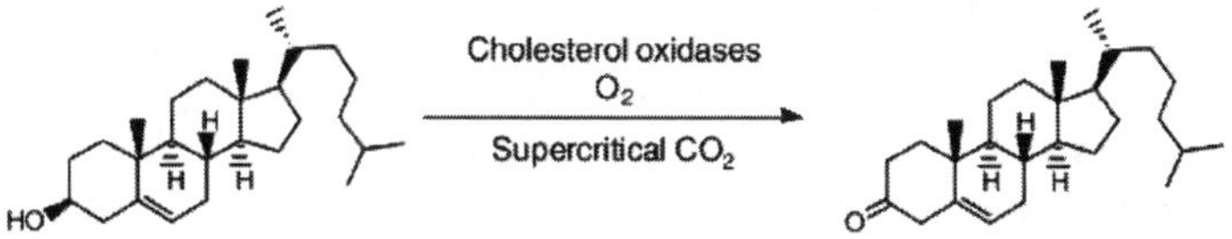 Cholesterol oxidases, O_2, Supercritical CO_2	176
4	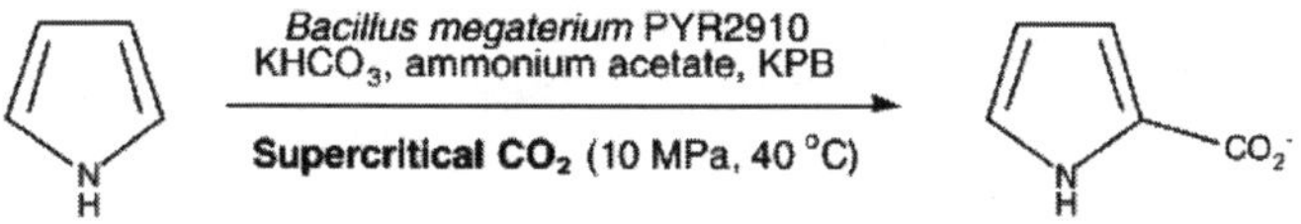 *Bacillus megaterium* PYR2910, $KHCO_3$, ammonium acetate, KPB, **Supercritical CO_2** (10 MPa, 40 °C) (product: pyrrole–CO_2^-)	177

Table 16. Enzyme catalyzed reduction of ketones in supercritical CO_2

Sr. No.	Substrate	Yield %	ee%	Configuration
1	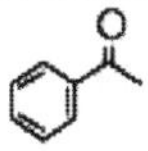	51	>99	S
2	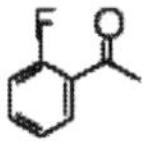	81	>99	S
3	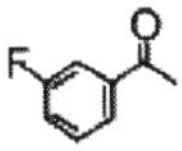	53	>99	S
4	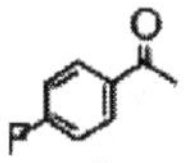	11	97	S
5	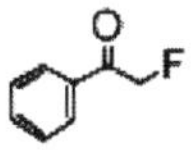	96	96	R
6	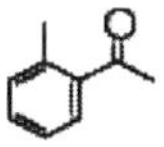	22	>99	S
7	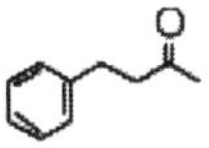	61	>98	S
8	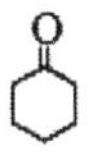	96	–	–

The advantage of alcohol dehydrogenase catalyzed reactions in supercritical CO_2 is the ease of the product isolation from CO_2. When an aqueous solvent is used, the product has to be extracted, but this is unnecessary when using supercritical CO_2. The whole resting cell instead of an isolated enzyme was used for the reduction, and thus no addition of an expensive co-enzyme was required. Very high enantioselectivities (>99% ee) were obtained for the reduction with the majority of the substrates tested (Table 16).

Synergism of Microwave Irradiation and Enzymes

Microwave irradiation has been used for a wide variety of purposes such as drying, moisture analysis, dissolution of geological materials and rapid hydrolysis of peptides and proteins.[181] Apart from these applications microwave is used by organic chemist for its utilization in organic synthesis. Accelerations by microwave have been observed for a wide range of organic reactions. Microwave irradiation is a very convenient, safe and rapid methodology.[182] The microwave heating process is however, fundamentally different from the heating process used conventionally. Microwave radiation has the effect on enzyme activity as a function of enzyme hydration.[183] They result from material–wave interactions leading to thermal effects (connected to dipolar and charge space polarization) and specific (purely nonthermal) effects.[184] The thermal gradients and flow of heat is the reverse of those in materials heated by conventional means. Enzyme catalyzed reactions are rather sluggish in nature; the synergism with microwave enhance the rates of reactions.[185-187] In a number of chemically catalyzed reactions microwave irradiation catapults the reactions rates by order of magnitude and the rate of enzymatic reactions is also enhanced.[188-191] Lipase has been extensively studied under microwave irradiation. Various types of reactions carried out using lipase are described in Table 17.

Table 17. Various lipase catalyzed reactions carried under microwave irradiation

Sr. No	Type of Reaction	Ref.
1	Esterification $HOOC(CH_2)_4COOH$ (Adipic acid) + C_4H_9OH (n-Butanol) $\xrightarrow{Lipase}$ $C_4H_9OOC(CH_2)_4COOH$ (Monobutyl adipate) $C_4H_9OOC(CH_2)_4COOH$ (Monobutyl adipate) + C_4H_9OH (n-Butanol) $\xrightarrow{Lipase}$ $C_4H_9OOC(CH_2)_4COOC_4H_9$ (Dibutyl adipate)	185
2	Transesterification H_3C–CO–CH_2–CO–OCH_3 (Ethyl acetoacetate) + ROH $\xrightarrow{Lipase}$ H_3C–CO–CH_2–CO–OR + CH_3OH	184
3	Epoxidation $CH_3(CH_2)_{10}COOH$ (Lauric acid) + H_2O_2 (Hydrogen peroxide) $\xrightarrow{Lipase}$ $CH_3(CH_2)_{10}COOOH$ (Perlauric acid) + H_2O (Water) Styrene (C_6H_5–CH=CH2) + $CH_3(CH_2)_{10}COOOH$ (Perlauric acid) $\longrightarrow$ Styreneoxide + $CH_3(CH_2)_{10}COOH$ (Lauric acid)	186

Conclusions

Enzyme catalysis is being vigorously applied in a variety of processes to make several fine chemicals and pharmaceuticals in aqueous and non-aqueous media including organic solvents, ionic liquids and supercritical fluids. This is an age of enantiopure drugs involving millions of dollars and enzyme catalysis will be most attractive in comparison with the traditional chemical catalysis. The synergism of microwave irradiation with enzyme catalysis is also being pursued vigorously in several laboratories for separation and synthesis of pharmaceuticals.

Acknowledgements

Thanks are due to the Department of Biotechnology, Govt. of India for support vide grant #BT/PR2942/PID/06/149/2002 on "Synthesis of Chiral Drugs through Biotransformation" and the Darbari Seth Endowment for the Chair Professorship to GDY.

References

1. U.S. Industry and Trade Outlook 2000, New York: The McGraw-Hill Companies, 2001:11.1-11.19.
2. Thayer AM. Biocatalysis. C & E News 2001; 79:27-34.
3. Straathof AJJ, Panke S, Schmid A. The production of fine chemicals by biotransformations. Curr Opin Biotechnol 2002; 13:548-556.
4. Schulze B, Broxterman R, Shoemaker H, Boesten W. Review of biocatalysis in production of chiral fine chemicals. Spec Chem 1998; 18:244-246.
5. Stinson SC. Chiral drugs. C & E News 2000; 78:55-77.
6. Kaplan DL, Dordick JS, Gross RA, Swift G. Enzymes in polymer science: an introduction. Am Chem Soc Symp Ser 1998; 684:2-16.
7. Paradkar VM, Dordick JS. Aqueous like activity of α-chymotrypsin dissolved in nearly anhydrous organic solvents. J Am Chem Soc 1994; 116:1009-1010.
8. Thomas SM, DiCosimo R, Nagarajan V. Biocatalysis: applications and potentials for chemical industry. Trends Biotechnol 2002; 6:238-242.
9. Bühler B, Witholt B, Hauer B, Schmid A. Characterization of xylene monooxygenase for mutistep biocatalysis. Appl Environ Microbial 2002; 68:560-568.
10. Fessner W-D, Helaine V. Biocatalytic synthesis of hydroxylated natural products using aldolases and related enzymes. Curr Opin Biotechnol 2001; 12:574-586.
11. Weissermel K, Arpe HJ. Components for polyamide, In: Industrial Organic Chemistry. Weinhem: VCLL, 1993:257.
12. Brzostowicz PC, Gibson KL, Thomas SM et al. Simultaneous identification of two cyclohexanone oxidation genes from a new Brevibacterium sp. using mRNA differential display. J Bacteriol 2000; 182:4241-4248.
13. Chen YC, Peoples OP, Walsh CT. Acinetobacter cyclohexanone monooxygenase: gene cloning and sequence determination. J Bacteriol 1988; 170:781-789.
14. Banerjee A. Stereoselective microbial Baeyer-Villiger oxidations. In: Patel RN, ed. Stereoselective Biocatalysis. Marcel-Dekkar Inc., 2000:867-876.
15. Kostichka K, Thomas SM, Gibson KJ et al. Cloning and characterization of a gene cluster for cyclodecanone oxidation in Rhodococcus rubber SC1. J Bacteriol 2001; 183:6478-6486.
16. Shumacher JD, Fakaussa RM. Degradation of allicyclic molecules by Rhodococcus rubber CD1. Appl Microbial Biotechnol 1999; 52:85-90.
17. Cheng Q et al. Genetic analysis of gene cluster for cyclohexanol oxidation in Acinetobacter NCIMB 9871. Eur J Biochem 2000; 60:1-7.
18. Bramucei MG et al. Microbial production of terephthalic acid and isophthalic acid. 2001; US patent no. 6187569.
19. Kobayashi M, Shimizu S. Nitrile hydrolases. Curr Opin Chem Biol 2000; 4:95-102.
20. Yamada H, Kobayashi M. Nitrile hydratase and its application to industrial production of acrylamide. Biosel Biotechnol Biochem 1996; 60:1391-1400.
21. Chassin C. A biotechnological process for the production of nicotinamide. Chim Oggi 1996; 14:9-12.
22. Robins KT, Nagasawa T. Process for preparing amides PCT Int. 1999; Appl. WO 9905306 A118.
23. Gavagan JE et al. A gram negative bacterium producing a heat stable nitrilase highly active on aliphatic dinitriles. Appl Microbial Biotechnol 1999; 52:654-659.
24. Cooling FB et al. Chemoenzymatic production of 1,5-dimethyl-2- piperidone, J Mol Catal B Enzym 2001; 11:295-306.
25. Gavagan JE et al. Chemoenzymatic production of lactams from aliphatic dinitriles. J Org Chem 1998; 63:4792-4801.

26. Breinig S, Schiltz E, Fuchs G. Genes involved in anaerobic metabolism of phenol in the bacterium Thaurea aromatica. J Bacteriol 2000; 182:5849-5863.
27. Aresta M et al. Enzymatic synthesis of 4-OH benzoic acid from phenol and CO_2: the first example of biotechnological application of a carboxyalse enzyme. Tetrahedron 1998; 54:8841-8846.
28. Laffend LA et al. Bioconversion of a fermentable carbon source to 1,3 propanediol by a single micro-organism. 1997; US patent no 5688276.
29. Drauz K, Waldmann H. Enzyme Catalysis in Organic Synthesis, a Comprehensive Handbook. Vol. II. VCH publications, 1995:688-689.
30. Bevinakatti HS, Banerji AA. Lipase catalysis in organic solvents: application to the synthesis of (R) and (S) Atenolol. J Org Chem 1992; 57(22):6003-6005.
31. Bevinakatti HS, Banerji AA, Newadkar RV. Resolution of secondary alcohols using lipase in diisopropyl ether. J Org Chem 1989; 54(10):2453-2455.
32. Ondetti MA, Rubin B, Cushman DW. Design of specific inhibitors of angiotensin-converting enzyme: new class of orally active antihypertensive agents. Science 1977; 196:441-444.
33. Ondetti MA, Cushman DW. Inhibition of rennin-angiotensin system. A new approach to the therapy of hypertension. J Med Chem 1981; 24(4):355-361.
34. Ondetti MA, Cushman DW. Inhibitors of angiotensin-converting enzyme for treatment of hypertension. Biochem Pharma 1980; 29:8171-1875.
35. Gu Qu-Ming, Reddy DR, Sih CJ. Bifunctional chiral synthons via biochemical methods. VIII. Optically active 3-arylthio-2-methylpropionic acids. Tetrahedron Lett 1986; 27(43):5203-5206.
36. Sakime A, Yuri K, Ryozo N, Hisao O. Structure and synthesis of angiotensin inhibitor. Eur Patent 0,172,614. 1986.
37. Patel RN, Howell JM, McNamee CG et al. Synthesis of chiral side chain of captopril using lipase. Biotechnol Appl Biochem 1992; 16:34-47.
38. Patel RN, Robinson RS, Szarka LJ. Stereoselective enzymic hydrolysis of 2-cyclohexyl and 2-phenyl-1,3-propanediol diacetate in biphasic systems. Appl Microbiol Biotechnol 1990; 34:10-14.
39. Almisick AV, Buddrus J, Honicke-Schmidt P et al. Enzymatic preparation of optically active cyanohydrin acetates. J Chem Soc Chem Commun 1989; 18:1391-1393.
40. Patel RN, Robinson RS, Szarka LJ et al. Stereospecific microbial reduction of 4,5-dihydro-4-(4-methoxyphenyl)-6-(trifluoromethyl-1H-1)-benzazepin-2-one. Enzyme Microb Technol 1991; 13(11):906-912.
41. Erzan F, Trani M, Lortie R. Selective esterification of racemic ibuprofen. Enzyme Engineering VII. Ann N Y Acad Sci 1995; 750:228-231.
42. Duan G, Ching CB, Lim E, Ang CH. Kinetic study of enantioselective esterification of ketoprofen with n-propanol catalyzed by lipase in an organic medium. Biotechnol Lett 1997; 19(11):1051-1055.
43. Barbara L-S, Wolfgang K, Kurt F. Chemoenzymatic synthesis of (R)- and (S)-2-hydroxy-4-phenylbutanoic acid via enantiocomplementary deracemisation of (±) -2-hydroxy-4-phenyl-3-butenoic acid using racemase-lipase two enzyme system. Synlett 2005; 12:1936-1938.
44. Palomo JM, Segura RL, Fuentes M et al. Unusual enzymatic resolution of (±)-glycidyl-butyrate for the production of (S)-glycidyl derivatives. Enzyme Microb Technol 2006; 38(3-4):429-435.
45. Harrison IT, Lewis B, Nelson P et al. Nonsteroidal anti-inflammatory agents. I. 6-substituted-2-naphthylacetic acids. J Med Chem 1970; 13(2):203-205.
46. Hayball PJ. Chirality and nonsteroidal anti-inflammatory drugs. Drugs 1996; 52:47-58.
47. Bhandarkar SV, Neau SH. Lipase catalyzed enantioselective esterification of flurbiprofen with n-butanol. Electronic J Biotechnol 2000:3(3).
48. Kim MG, Lee EG, Chung BH. Improved enantioselectivity by candida rugosa lipase towards ketoprofen ethyl ester by a simple two-step treatment. Process Biochem 2000; 35(9):977-982.
49. Lee K-W, Bae H-A, Shin G-S, Lee Y-H. Purification and catalytic properties of novel enantioselective lipase from acinetobacter Sp. ES-1 for hydrolysis of (S)-ketoprofen ethyl ester. Enzyme Microb Technol 2006; 38(3-4):443-448.
50. Ong AL, Kamaruddin AH, Bhatia S. Current technologies for the production of (S)-Ketoprofen: Process perspective. Process Biochem 2005; 40(11):3526-3535.
51. Steenkamp L, Brady D. Screening of commercial enzymes for the enantioselective hydrolysis of (R,S) naproxen ester. Enzyme Microb Technol 2003; 32 (3-4):472-477.
52. Franz E, Walter GB, Steffen O. Stereoselective hydrolysis of racemic naproxen amide with Rhodococcus Erythropolis. Tetrahedron: Asymmetry 1997; 8(16):2749-2755.
53. Spencer JL, Flynn EH, Roeske RW et al. Chemistry of cephalosporin antibiotics VII. Synthesis of cephaloglycin and some homologs. J Med Chem 1966; 9(5):746-750.
54. Ohi N, Aoki B, Shinozaki T et al. Semisynthetic β-lactam antibiotics I. Synthesis and antibacterial activity of new ureidopenicillin derivatives having catechol moieties. J Antibiot 1986; 39:230-241.

55. Choi WG, Lee SB, Ryu DDY. Cephalexin synthesis by partially purified and immobilized enzymes. Biotechnol Bioeng 1981; 23(2):361-371.
56. Takahashi T, Yamazaki Y, Kato K, Isono M. Enzymatic synthesis of cephalosporins. J Am Chem Soc 1972; 94(11):4035-4037.
57. Bruggink A, Roos EC, Vroom E. Penicillin acylase in the industrial production of β-lactam antibiotics. Org Process Res Dev 1998; 2(2):128-133.
58. Shaw S, Shyu J, Hsieh Y, Yeh H. Enzymatic synthesis of cephalothin by penicillin G acylase. Enzyme Microb Technol 2000; 26(2-4):142-151.
59. Giordano RC, Ribeiro MPA, Giordano RLC. Kinetics of H-lactam antibiotics synthesis by penicillin G acylase (PGA) from the viewpoint of the industrial enzymatic reactor optimization. Biotechnol Adv 2006; 24:27-41.
60. Hahn FE. In: Gottlieb D, Shaw PD, eds. Antibiotics. Vol. 1. Heidelberg: Springer-Verlag, 1967:308.
61. Pestka S. In: Corcoran JW, Hahn FE, eds. Antibiotics. Vol. 3. Heidelberg: Springer-Verlag, 1975:370.
62. Chenevert R, Thiboutot S. Synthesis of chloramphenicol via an enzymatic enantioselective hydrolysis. Synthesis 1989; 19(6):444-446.
63. Roach PL et al. The crystal structure of isopenicillin N synthase, first of a new structural family of enzymes. Nature 1995; 375:700-704.
64. Valegard K et al. Structure of a cephalosporin synthase. Nature 1998; 394:805-809.
65. Que L. One motif-many different reactions. Nature Struct Biol 2000; 7:182-184.
66. SinhaRoy R, Milne J, Belshaw P et al. Oxazole and thiazole peptide biosynthesis. Nat Prod Rep 1999; 16:249-263.
67. Lewis RJ, Tsai FT, Wigley DB. Molecular mechanism of drug inhibition of DNA gyrase. BioEssays 1996; 18:661-671.
68. Robertson DW, Krushinski JH, Fuller RW, Leander JD. Absolute configurations and pharmacological activities of the optical isomers of fluoxetine, a selective serotonin-uptake inhibitor. J Med Chem 1988; 31(7):1412-1417.
69. Ader U, Andersch P, Berger M et al. Hydrolases in organic synthesis: preparation of enantiomerically pure compounds. Indian J Chem 1993; 32B:145-150.
70. Kumar A, Dilip H, Dik SY. A new chemoenzymatic enantioselective synthesis of R-(-)-Tomoxetine and (R) and (S)-Fluoxetine. Tetrahedron Lett 1991; 32(16):1901-1904.
71. Hanson RL, Parker WL, Brzozowski DB et al. Preparation of (R) and (S)-6-hydroxybuspirone by enzymatic resolution or hydroxylation. Tetrahedron Asymmetry 2005; 16(16):2711-2716.
72. Patel RN, Chu L, Nanduri V et al. Enantioselective microbial reduction of 6-oxo-8-[4-[4-(2-pyrimidinyl)-1-piperazinyl]butyl]-8-azaspiro[4,5]-decane-7,9-dione. Tetra Asymm 2005; 16(16):2778-2783.
73. Process for production of (+) 2-amino-1-butanol. Japan Patent 1,469,014. Sept. 28, 1973, Denki Kagaku Kogyo Kabushiki Kaisha.
74. Yadav GD, Joshi SS, Lathi PS. Enzymatic synthesis of isoniazid in non-aqueous medium. Enzyme Microb Technol 2005; 36:217-222.
75. Magri NF, Kingston DGI. Modified Taxols 4. Synthesis and biological activity of taxols modified in the Side Chain. J Nat Prod 1988:298-306.
76. Fleming PE, Knaggs AR, He Xian-Guo et al. Biosynthesis of taxoids. Mode of attachment of the taxol side chain. J Am Chem Soc 1994; 116(9):4137-4138.
77. Lythgoe B. The Taxus Alkaloids. In: Manske RHF, ed. The Alkaloids Chemistry and Physiology. Vol. 10. New York: Academic Press, 1968:597-626.
78. Schiff PB, Fant J, Horwitz SB. Promotion of microtubule assembly in vitro by taxol. Nature 1979; 277:665-667.
79. Holton RA, Juo RR, Kim HB et al. A synthesis of taxus. J Am Chem Soc 1988; 110(19):6558-6560.
80. Denis JN, Greene AE, Aarao SA, Luck MJ. Efficient enantioselective synthesis of the taxol side chain. Org Chem 1986; 51:46-50.
81. Christen AA, Gibson DM, John B. Production of taxol and taxol-like compounds with taxus brevifolia callus cell culture. U S Patent 5,019,504, May 28, 1991.
82. Ojima I, Habus I, Zhao M et al. New and efficient approaches to the synthesis of taxol and its C-13 side chain analogs by means of β-lactam synthon method. Tetrahedron 1992; 48(34):6985-7012.
83. Patel RN, Banerjee A, Szarka LJ. Biocatalytic synthesis of some chiral pharmaceutical intermediates by lipases. J Am Oil Chem Soc 1996; 73(11):1363-1375.
84. Yasohara Y, Miyamoto K, Kizaki N et al. A practical chemoenzymatic synthesis of a key intermediate of antifungal agents. Tetrahedron Lett 2001; 2(19):3331-3333.
85. Philips RS, May SW. Enzymatic sulphur oxygenation reactions. Enzyme Microb Technol 1981; 3(1):9-18.
86. Holt R, Lindberg P, Reeve C, Taylor S. Enantioselective preparation of pharmaceutically active sulfoxides by biooxidation. June 6, 1996:WO 9,617,076.

87. Fabio P, Krzysztof O, Michel T. Bienzymatic synthesis of chiral heteroaryl-methyl-sulfoxides. Tetrahedron Asymmetry 2005; 16(16):2681-2683.
88. Demir AS, Hamamci H, Doganel F, Ozgul E. Chemoenzymatic synthesis of 1S,2 R -1-amino-2-indanol, a key intermediate of HIV protease inhibitor, indinavir. J Mol Catal B: Enzymatic 2000; 9:157-161.
89. Kamal A, Chouhan G. Chemoenzymatic synthesis of calcilytic agent nps-2143 employing a lipase mediated resolution protocol. Tetra Asymm 2005; 16(16):2784-2789.
90. Patel RN, Banerjee A, Pendri YR et al. Preparation of a chiral synthon for an HBV inhibitor: enzymatic asymmetric hydrolysis of (1α,2β,3α)-2-(benzyloxymethyl)cyclopent-4-ene-1,3-diol diacetate and enzymatic asymmetric acetylation of (1α,2β,3α)-2-(benzyloxymethyl)cyclopent-4-ene-1,3-diol. Tetra Asymm 2006; 17(2):175-178.
91. Willis MC. Enantioselective desymmetrization. J Chem Soc Perkin Trans 1 1999; 13:1765-1784.
92. Schoffers E, Golebiowski A, Johnson CR. Enantioselective synthesis through enzymatic asymmetrization. Tetrahedron 1996; 52:3769-3826.
93. Davoli P, Caselli E, Bucciarelli M et al. Lipase catalysed resolution and desymmetrization of 2-hydroxymethylaziridines. J Chem Soc Perkin Trans1 2002; 17:1948-1953.
94. Chenevert R, Dickmann M. Enzymatic route to chiral, nonracemic cis 2,6- and cis, cis 2,4,6- substituted piperidines. Synthesis of (+) dihydropinidine and dendrobate alkaloid (+) -241D. J Org Chem 1996; 61:3332-3341.
95. Matsumoto T, Konegawa T, Yamaguchi H et al. Lipase catalysed asymmetrization of diacetate of meso-2-(2-propynyl)cyclohexane-1,2,3-triol toward the total synthesis of aquayamycin. Synlett 2001; 10:1650-1652.
96. Toyama K, Iguchi S, Sakazaki H et al. Convenient route to both enantiomerically pure forms of trans-4,5-dihydroxy-2-cyclopenten-1-one: efficient synthesis of the neocarzino statin chromophore core. Bull Chem Soc Jpn 2001; 74:997-1008.
97. Hilpert H, Wirz B. Novel versatile approach to an enantiopure 19-nor, des-C,D vitamin D_3 derivative. Tetrahedron 2001; 57:681-694.
98 Ohta H. Stereochemistry of enzymic hydrolysis of nitriles. Chimia 1996; 50:434-436.
99. Yokoyama M, Sugai T, Ohta H. Asymmetric hydrolysis of a disubstituted malononitrile by the aid of a microorganism. Tetrahedron Asymmetry 1993; 4:1081-1084.
100. Osby JA, Parratt JS, Turner NJ. Enzymic hydrolysis of prochiral dinitriles. Tetrahedron Asymmetry 1992; 12:1547-1550.
101. Servi S A. Baker's yeast as a reagent in organic synthesis. Synthesis 1990; 1:1-25. b. Csuk R, Glanzer BI. Bakers yeast mediated transformations in organic chemistry. Chem Rev 1991; 91:49-97.
102. Matsuda T, Harada T, Nakajima N et al. Two classes of enzymes of opposite stereochemistry in an organism: one for fluorinated and another for nonfluorinated substrates. J Org Chem 2000; 65:157-163.
103. Matsuda T, Nakajima Y, Harada T, Nakamura K. Asymmetric reduction of simple aliphatic ketones with dried cells of Geotrichum candidum. Tetrahedron Asymmetry 2002; 13:971-974.
104. Stampfer W, Kosjek B, Moitzi C et al. Biocatalytic asymmetric hydrogen transfer. Angew Chem Int Ed 2002; 41:1014-1017.
105. Goswami A, Bezbaruah RL, Goswami J et al. Microbial reduction of ω-bromoacetophenones in the presence of surfactants. Tetrahedron Asymmetry 2000; 11:3701-3709.
106. Goswami J, Bezbaruah RL, Goswami A, Borthakur N. A convenient stereoselective synthesis of (R)-(−)-denopamine and (R)-(−)-salmeterol. Tetrahedron Asymmetry 2001; 12:3343-3348.
107. Yadav JS, Nanda S, Reddy PT, Rao AB. Effecient enantioselective reduction of ketones with Daucus carota root. J Org Chem 2002; 67:3900-3903.
108. Yadav JS, Reddy PT, Nanda S, Rao AB. A facile synthesis of (R)-(−)-2-azido-1-arylethanols from 2-azido-1-arylketones using baker's yeast. Tetrahedron Asymmetry 2001; 12:63-67.
109. Burns MP, Wong JW. US patent 6451587 2002. CA 2002; 137:246600h.
110. Fuhshuku K-i,Tomita M, Sugai T. Enantiomerically pure octahydronaphthalenone and octahydroindenone: elaboration of the substrate overcame the specificity of yeast mediated reduction. Adv Synth Catal 2003; 345:766-774.
111. Ishihara K, Yamaguchi H, Hamada H et al. Stereocontrolled reduction of α-keto esters with thermophilic actinomycete, Streptomyces thermocyaneoviolaceus IFO 14271. J Mol Catal B: Enzymatic 2000; 10:429-434.
112. Patel RN, Chu L, Chidambaram R et al. Enantioselective microbial reduction of 2-oxo-2-(1', 2', 3', 4'-tetrahydro-1', 1', 4', 4'-tetramethyl-6'-naphthalenyl)acetic acid and its ethyl ester Tetrahedron Asymmetry 2002; 13:349-355.
113. Schubert T, Hummel W, Kula MR, Muller M. Enantioselective synthesis of both enantiomers of various propargylic alcohols by use of two oxidoreductases. Eur J Org Chem 2001; 22:4181-4187.
114. Wolberg M, Hummel W, Wandrey C, Muller M. Highly regio- and enantiselective reduction of 3,5-dioxocaboxylates. Angew Chem Int Ed 2000; 39:4306-4308.

115. Molinari F, Gandolfi R, Villa R et al. Enantioselective oxidation of prochiral 2-methyl-1,3-propandiol by Acetobacter pasteurianus Tetrahedron Asymmetry 2003; 14:2041-2043.
116. Fugati C, Serra S. Baker's yeast medaietd enantioselective synthesis of the bisabolane sesquiterpenes (+)-curcuphenol, (+)-xanthorrhizol, (-)-curcuqninone and (+)-curcuhydroquinone. J Chem Soc Per Trans1 2000; 22:3758-3764.
117. Izquierdo I, Plaza MT, Rodriguez M, Tamaya J. Chiral building-blocks by chemoenzymatic desymmetrization of 2-ethyl-1,3-propanediol for the preparation of biologically active natural products. Tetrahedron Asymmetry 1999; 10:449-455.
118. Neri C, Williams JMJ. New routes to chiral evans auxiliaries by enzymatic desymmetrisation and resolution strategies. Adv Synth Catal 2003; 345:835-848.
119. Takabe K, Iida Y, Hiyoshi H et al. Reverse enantioselectivity in the lipase-catalyzed desymmetrization of prochiral 2-carbamoylmethyl-1,3-propanediol derivatives. Tetrahedron Asymmetry 2000; 11:4825-4829.
120. Chenevert R, Courchesne G. Synthesis of (S)-α-tocotrienol via an enzymatic desymmetrization of an achiral chroman derivative. Tetrahedron Lett 2002; 43:7971-7973.
121. Kirihara M, Kawasaki M, Takuwa T et al. Efficient synthesis of (R)- and (S)-1-amino-2,2-difluorocyclopropanecarboxylic acid via lipase-catalyzed desymmetrization of prochiral precursors. Tetrahedron Asymmetry 2003; 14:1753-1761.
122. Chenevert R, Courchesne G, Caron D. Chemoenzymatic enantioselective synthesis of the polypropionate acid moiety of dolabriferol. Tetrahedron Asymmetry 2003; 14:2567-2571.
123. Chenevert R, Rose YS. Enzymatic desymmetrization of a meso polyol corresponding to the C(19)-C(27) segment of rifamycin S. J Org Chem 2000; 65:1707-1709.
124. Ohrlein R, Baisch G. Chemo-enzymatic approach to statin side-chain building blocks. Adv Synth Catal 2003; 345:713-715.
125. Wu Z-L, Li Z-Y. Enantioselective biotransformation of α,α-disubstituted dinitriles to the corresponding 2-cyanoacetamides using Rhodococcus sp. CGMCC 0497. Tetrahedron Asymmetry 2003; 14:2133-2142.
126. Nakamura K, Takenaka K, Fujii M, Ida Y. Asymmetric synthesis of both enantiomers of secondary alcohols by reduction with a single microbe. Tetrahedron Lett 2002; 43:3629-3631.
127. Barbieri C, Caruso E, D'Arrigo P et al. Chemo-enzymatic synthesis of (R)- and (S)-3,4-dichlorophenylbutanolide intermediate in the synthesis of sertraline Tetrahedron Asymmetry 1999; 10:3931-3937.
128. Wei ZL, Lin GQ, Li ZY. Microbial transforamtion of 2-hydroxy and 2-acetoxy ketones with Geotrichum sp. Bioorg Med Chem 2000; 8:1129-1137.
129. Ward RS. Dynamic kinetic resolution. Tetrahedron Asymmetry 1995; 6:1475-1490.
130. Stürmer R. Enzymes and transition metal complexes in tandem—a new concept for dynamic kinetic resolution. Angew Chem Int Ed Engl 1997; 36:1173-1174.
131. El Gihani MT, Williams JMJ. Dynamic kinetic resolution. Curr Opin Biotechnol 1999; 3:11-15.
132. Azerad R, Buisson D. Dynamic resolution and stereoinversion of secondary alcohols by chem-enzymatic processes. Curr Opin Biotechnol 2000; 11:565-571.
133. Huerta FF, Minidis ABE, Bäckvall JE. Racemisation in asymmetric synthesis. Dynamic kinetic resolution and related processes in enzyme and metal catalysis. Chem Soc Rev 2001; 30:321-331.
134. Kim M-J, Ahn Y, Park J. Dynamic kinetic resolutions and asymmetric transformations by enzymes coupled with metal catalysis. Curr Opin Biotechnol 2002; 13(6):578-587.
135. Reetz MT, Schimossek K. Lipase-catalyzed dynamic kinetic resolution of chiral amines: use of palladium as the racemization catalyst. Chimia 1996; 50:668-669.
136. Dinh PM, Howarth JA, Hudnott AR et al. Catalytic racemisation of alcohols: application to enzymatic resolution reactions. Tetrahedron Lett 1996; 37:7623-7626.
137. Huerta FF, Bäckvall J-E. Enantioselective synthesis of (β)-hydroxy acid derivatives via a one-pot aldol reaction-dynamic kinetic resolution. Org Lett 2001; 3:1209-1212.
138. Koh JH, Jeong HM, Park J. Efficient catalytic racemization of secondary alcohols. Tetrahedron Lett 1998; 39:5545-5548.
139. Martin-Matute B, Edin M, Bogar K et al. Combined ruthenium(II) and lipase catalysis for efficient dynamic kinetic resolution of secondary alcohols. Insight into the racemization mechanism. J Am Chem Soc 2005; 127(24):8817-8825.
140. Lin CN, Tsai SW. Dynamic kinetic resolution of suprofen thioester via coupled trioctylamine and lipase catalysis. Biotechnol Bioeng 2000; 69(1):31-38.
141. Wuyts S, De Temmerman K, De Vos D, Jacobs P. A zeolite-enzyme combination for biphasic dynamic kinetic resolution of benzylic alcohols. Chem Commun (Camb) 2003; 15:1928-1929.
142 Schnell B, Faber K, Kroutil W. Enzymatic racemisation and its application to synthetic biotransformations. Adv Synth Catal 2003; 345:653-666.

143. Paizs C, Tähtinen P, Lundell K et al. Preparation of novel phenylfuran-based cyanohydrin esters: lipase-catalysed kinetic and dynamic resolution. Tetrahedron Asymmetry 2003; 14:1895-1904.
144. Pàmies O, Bäckvall J-E. Combined metal catalysis and biocatalysis for an efficient deracemization process. Curr Opin Biotechnol 2003; 14:407-413.
145. Turner NJ. Enzyme catalysed deracemisation and dynamic kinetic resolution reactions. Curr Opin Chem Biology 2004; 8:114-119.
146. a. Persson BA, Larsson ALE, LeRay M, Bakvall JE. Ruthenium and enzyme catalysed dynamic kinetic resolution of secondary alcohols. J Am Chem Soc 1999; 121:1645-1650. b. Choi JH, Kim YH, Nam SH et al. Aminocyclopentadienyl ruthenium chloride: catalytic racemization and dynamic kinetic resolution of alcohols at ambient temperature. Angew Chem Int Ed Engl 2002; 41:2373-2376.
147. Jung HM, Koh JH, Kim MJ, Park J. Practical ruthenium/lipase-catalyzed asymmetric transformations of ketones and enol acetates to chiral acetates. Org Lett 2000; 2:2487-2490.
148. Pamies O, Backwall JE. Enzymatic kinetic resolution and chemoenzymatic dynamic kinetic resolution of delta-hydroxy esters. An efficient route to chiral delta-lactones. J Org Chem 2002; 67:1261-1265.
149. Gutierrez MC, Furatoss R, Alphand V. Microbiological transformations 60. Enantioconvergent Baeyer-Villiger oxidation via a combined whole cells and ionic exchange resin catalysed dynamic kinetic process. Adv Syn Catal 2005; 347(7+8):1051-1059.
150. Yang Zhen, Pan Wubin. Ionic liquids: Green solvents for nonaqueous biocatalysis. Enzyme Microb Technol 2005; 37:19-28.
151. Eckstein M, Sesing M, Kragl U, Adlercreutz P. At low water activity α-chymotrypsin is more active in an ionic liquid than in non-ionic organic solvents. Biotehnol Lett 2002; 24:867-872.
152. Kaar JL, Jesionowski AM, Berberich JA et al. Impact of ionic liquid physical properties on lipase activity and stability. J Am Chem Soc 2003;125:4125-4131.
153. Schöfer SH, Kaftzik N, Wasserscheid P, Kragl U. Enzyme catalysis in ionic liquids: lipase-catalysed kinetic resolution of 1- phenyl ethanol with improved enantioselectivity. CheCommun 2001:425-426.
154. Zhao H, Luo RG, Malhotra SV. Kinetic study on the enzymatic resolution of homophenylalanine ester using ionic liquids. Biotechnol Prog 2003;19:1016-1018.
155. Park S, Kazlauskas R. Improved preparation and use of room temperature ionic liquids in lipase-catalyzed enantio- and regioselective acylations. J Org Chem 2001; 66:8395-8401.
156. Eckstein M, Villela FM, Liese A, Kragl U. Use of an ionic liquid in a two-phase system to improve an alcohol dehydrogenase catalysed reduction. Chem Commun (Camb) 2004:1084-1085.
157. Erbeldinger M, Mesian AJ, Russell AJ. Enzymatic catalysis of formation of Z-aspartame in ionic liquid—an alternative to enzymatic catalysis in organic solvents. Biotechnol Prog 2000; 16:1129-1131.
158. Laszlo JA, Compton DL. Comparison of peroxidase activities of hemin, cytochrome c and microperoxidase-11 in molecular solvents and imidazolium-based ionic liquids. J Mol Catal B Enzym 2002; 18:109-120.
159. Kaftzik N, Wasserscheid P, Kragl U. Use of ionic liquids to increase the yield and enzyme stability in the galactosidase catalysed synthesis of N-acetyllactosamine. Org Proc Res Dev 2002; 6:553-557.
160. Pfruender H, Amidjojo M, Kragl U, Weuster-Botz D. Efficient whole-cell biotransformation in a biphasic ionic liquid/water system. Angew Chem Int Ed Engl 2004; 43:4529-4531.
161. Sheldon R.A, Lau RM, Sorgedrager MJ. Rantwijk Fv, Seddon KR. Biocatalysis in ionic liquids. Green Chemistry 2002; 4:147-151
162. Erbeldinger M, Mesiano M, Russell AJ. Enzymatic catalysis of formation of Z-aspartame in ionic liquid—an alternative to enzymatic catalysis in organic solvents. Biotechnol Prog 2000;16:1129-1131.
163. Park S, Kazlauskas RJ. Improved preparation and use of room temperature ionic liquids in lipase-catalyzed enantio- and regioselective acylations. J Org Chem 2001; 66:8395-8401.
164. de Zoete M.C. PhD thesis. Delft, The Netherlands: Delft University of Technology, 1995.
165. de Zoete M, Kock-van Dalen AC, Rantwijk F, Sheldon RA. Ester ammoniolysis: A new enzymatic reaction. Chem Commun 1993:1831-1832.
166. Howarth J, James P, Dai J. Immobilized baker's yeast reduction of ketones in an ionic liquid, [bmim]PF6 and water mix. Tetrahedron Lett 2001; 42:7517-7519.
167. McLeod R, Prosser H, Fiskentscher L et al. Asymmetric reductions. 12. Stereoselective ketone reductions by fermenting yeast. Biochemistry 1964; 3:838-846.
168. Fauve A, Verchambre H. Microbiological reduction of acyclic β-diketones. J Org Chem 1988; 53:5215-5219.
169. Jayasinghe LY, Smallridge AJ, Trewhella MA. The yeast mediated reduction of ethyl acetoacetate in petroleum ether. Tetrahedron Lett 1993; 34:3949-3950.
170. Nakamura K, Inoue K, Ushio K et al. Stereochemical control on yeast reduction of alpha.-keto esters. Reduction by immobilized bakers' yeast in hexane. J Org Chem 1988; 53:2589-2593.
171. Matsuda T, Watanabe K, Harada T, Nakamura K. Enzymatic reactions in supercritical CO_2: carboxylation, asymmetric reduction and esterification. Catalysis Today 2004; 96:103-111.

172. Mesiano AJ, Beckman EJ, Russell AJ. Supercritical biocatalysis. Chem Rev 1999; 99:623-634.
173. Yoon S-H, Miyawaki O, Park K-H, Nakamura K. Transesterification between triolein and ethylbehenate by immobilized lipase in supercritical carbon dioxide. J Ferment Bioeng 1996; 82:334-340.
174. Randolph TW, Blanch HW, Prausnitz JM, Wilke CR. Enzymatic catalysis in a supercritical fluid. Biotechnol Lett 1985; 7:325-328.
175. Mori T, Okahata Y. Effective biocatalytic transglycosylation in a supercritical fluid using a lipid-coated enzyme. Chem Commun 1998:2215-2216.
176. Mori T, Kobayashi A, Okahata Y. Biocatalytic esterification in supercritical carbon dioxide by using a lipid-coated lipase. Chem Lett 1998:921-922.
177. Matsuda T, Ohashi Y, Harada T et al. Conversion of pyrrole to pyrrole-2-carboxylate by cells of B. megaterium in supercritical CO_2. Chem Commun 2001:2194-2195.
178. Matsuda T, Harada T, Nakajima N et al. Two classes of enzymes of opposite stereochemistry in an organism: one for fluorinated and another for nonfluorinated substrates. J Org Chem 2000; 65:157-163.
179. Nakamura K, Matsuda T. Asymmetric reduction of ketones by the acetone powder of Geotrichum candidum. J Org Chem 1998; 63:8957-8964.
180. Nakamura K, Inoue Y, Matsuda T, Misawa I. Stereoselctive oxidation reduction by immobilized Geotrichum candidum in an organic solvent. J Chem Soc Perkin Trans I 1999:2397-2402.
181. Wathey B, Tierney J, Lidström P, Westman J. The impact of microwave-assisted organic chemistry on drug discovery. Drug Discovery Today 2002; 7(6):373-380.
182. Parker M, Besson T, Lamare S, Legoy M. Microwave radiation can increase the rate of enzyme-catalysed reactions in organic media. Tetrahedron Lett 1996; 37(46):8383-8386
183. Loupy A, Perreux L, Marion L et al. Reactivity and selectivity under microwaves in organic chemistry. Relation with medium effects and reaction mechanisms. Pure Appl Chem 2001; 73(1):161-166.
184. Yadav GD, Lathi PS. Synergism between microwave and enzyme catalysis in intensification of reactions and selectivities: transesterification of methyl acetoacetate with alcohols. J Mol Cat A: Chem 2004b; 223:51-56.
185. Yadav GD, Lathi PS. Synergism of microwave and immobilized enzyme catalysis in synthesis of adipic acid esters in non-aqueous media. Synth Comm 2005b; 35(12):1699-1705
186. Yadav GD, Borkar IV. Kinetic modeling of microwave assisted chemo-enzymatic epoxidation of styrene to styrene oxide. AIChEJ 2006; 52:1235-1247.
187. Bradoo S, Rathi P, Saxena R, Gupta J. Microwave-assisted rapid characterization of lipase selectivities. Biochem Biophys Methods 2002; 51:115-200.
188. Lin G, Lin WY. Microwave-promoted lipase-catalyzed reactions. Tetrahedron lett 1998; 39:4333-4336.
189. Vacek M, Zarevucka M, Wimmer Z et al. Selective enzymic esterification of free fatty acids with n-butanol under microwave irradiation and under classical heating. Biotechnol Lett 2000; 22:1565-1570.
190. Yadav GD, Lathi PS. Intensification of enzymatic synthesis of propylene glycol monolaurate from 1,2-propanediol and lauric acid under microwave irradiation: Kinetics of forward and reverse reactions. Enzyme Microb Technol 2006; 38(6):814-820.
191. Roy I., Gupta MN. Non-thermal effects of microwaves on protease-catalyzed esterification and transesterification. Tetrahedron 2003; 59(29):5431-5436.

Chapter 9

Utilization of Enzymes and Enzyme Mixtures in Abatement of Carbon Dioxide Pollution at Source

Mabel Algeciras* and Sanjoy K. Bhattacharya*

Abstract

Carbon dioxide is a major gaseous pollutant released as a result of burning carbonaceous material and it has been implicated in global warming. The utilization of enzyme mixtures for continuous fixation of carbon dioxide from emission sources has been reviewed. Two novel approaches have been recently developed for continuous fixation of carbon dioxide in emission streams. The first is to capture the gaseous carbon dioxide from the emission stream and the second is a process to convert the captured carbon dioxide into concatenated carbon compounds. Capture of carbon dioxide utilizes immobilized carbonic analydrase columns. Fixation of carbon dioxide is achieved with a cohort of enzymes that are divided into three modules. In the first module, carbon dioxide and ribulose bisphosphate (RuBP)is converted into 3-phophoglycerate (3PGA) catalyzed by Rubisco. In the second module, adenosine triphosphate (ATP) is generated using adenosine diphosphate (ADP) and inorganic phosphate is catalyzed by immobilized F0F1 ATPase. In the third module, a cohort of enzymes converts 3PGA into RuBP for recycling in the first module. Excess 3PGA generated due to carbon dioxide fixation is extracted. This complex process therefore utilizes a number of enzymes and results in continuous fixation of the carbon dioxide generating concatenated compound 3-phosphoglycerate, the starting material for further enzymatic conversions resulting in other useful compounds of commercial value. An overview of these processes and their utilization of enzyme mixtures is presented here.

Introduction

The anthropogenic emission of gases in the atmosphere has unintended consequences, which have been well documented and have caused global concern.[1-3] Each year, processing of fossil fuel generates more than 6.5 billion tons of carbon dioxide emitted in the atmosphere. Hydrocarbons, the constituents of fossil fuel, are chemically concatenated carbon chains with hydrogen as only heteroatoms attached to them. They are only a means of obtaining energy stored in their bonds and not energy by themselves; so are the other chemicals such as hydrogen which could be used as fuel. Hydrocarbons have been used as fuel for a very long time and thus it is probably reasonable to assume that most facets of problems associated with their use at all levels are better understood than other equivalent means of obtaining energy. Many other proposed means of chemical energy including hydrogen appear to be convenient, but lack the equivalent long experience that hydrocarbon enjoys and the global consequences of their use remains uncertain.[4-6] New means of chemical

*Sanjoy K. Bhattacharya and Mabel Algeciras—Bascom Palmer Eye Institute, McKnight Bldg., 1638 NW 10th Avenue, University of Miami, Miami, Florida 33136, U.S.A.
Emails: sbhattacharya@med.miami.edu; malgeciras@med.miami.edu

Enzyme Mixtures and Complex Biosynthesis, edited by Sanjoy K. Bhattacharya.

energy would require replacing existing equipment that uses fossil fuel, such as internal combustion engines. The activities for replacement itself will add a great degree of environmental pollution.

For polluting units, the abatement of carbon dioxide emission does not bring economic benefit for them and thus is viewed as a regulatory burden. It is clear that new and unique ways that enable fixation of carbon dioxide at source is necessary, but that is yet far away from practice, either at industries or in the bioengineering academia. The first step of a neu technology—one that has the potential to be a frontrunner for fuel recycling with the ability to abate carbon dioxide pollution at the source and generate chemical byproducts that have high comercial value—is continuous fixation of carbon dioxide at source. Enzymes are not even considered for combating industrial gas pollutants, whether carbon dioxide or any other gaseous material, because of the dogma prevalent in both industry and academia that sophisticated enzymatic methods cannot work in an industrial setup dealing with large-scale gas emission.

A sizable number of industrial processes emit CO_2 into the atmosphere. The current best approaches are pumping the gas into the sea (for operations based around sea-shores), which causes algal bloom and other unintended long-term consequences. For a short term, CO_2 can be fixed in the terminals of oxides:

$$CaO + CO_2 = CaCO_3$$

The terminal fixation of CO_2 is as easily reversed as fixed and does not hold the carbon in a bound state for a long enough period in the global carbon cycle. The only solution for holding carbon for a long term is concatenated carbon compound formation.[7] De novo formation of concatenated carbon compounds is very energy intensive and requires environmentally unfriendly reaction conditions. Synthesis of concatenated carbon compounds using CO_2 and a concatenated acceptor compound is possible. Methods to regenerate the acceptor from product of fixation is lacking a number of synthetic chemical compounds that can be used for such reactions, thus prohibiting their use. Lack of a regeneration scheme for any potential chemical acceptor renders the biological fixation schemes attractive. The plants use 5-carbon D-ribulose-1,5-bisphosphate, a concatenated carbon compound for fixation of CO_2, resulting in 3-phosphoglycerate (3PGA), a 3-carbon concatenated compound. The 3PGA can be used to regenerate RuBP but in the living world all known regeneration schemes proceed with loss of fixed carbon as CO_2. A scheme enabling regeneration of RuBP using 3PGA, but without loss of CO_2 does not exist in nature, but it was engineered.[8] The RuBP regeneration scheme needs ATP and NADH. A device harboring uniformly oriented immobilized F0F1 ATPase enables ATP synthesis using electrochemical potential gradient using solar energy for permissible hours and alternatively can be operated with stored electrical energy.[8,9] A similar efficient NADH recycling system helps in generating cofactor NADH. For continuous CO_2 fixation, generating 3PGA from RuBP maintenance of Rubisco in an insoluble state is necessary, for this reason immobilization of Rubisco was performed and shown to retain its activity.[7] Use of soluble or immobilized Rubisco directly with gas from emission stream is not possible, which will result in its irreversible inactivation. For capture of CO_2 from emission stream to be fed to Rubisco as carbonate solution, reactors harboring immobilized carbonic anhydrase were constructed.[10-12] Currently the scope of improvement remains for continuous CO_2 fixation involves: (1) devising an improved carbonic anhydrase capture system that will tolerate impurities including particulate matter, fluctuations in CO_2 concentration and temperature, and that is adaptable for coupling to Rubisco reactors. (2) Constructing adaptable Rubisco reactors that can withstand varying temperatures and fluctuation in soluble CO_2, pH. (3) Development of methods that enable creating complexes of functionally interacting enzymes leading to improved RuBP regeneration reactors. (4) Development of an improved reactor system, enabling recycling of NADH required for RuBP synthesis.

Capture of CO_2 at Emission Source

As shown in Figure 1, novel reactors have been made to capture CO_2 directly from the emission stream. These reactors were demonstrated to capture CO_2 from simulated emission. These reactors harbors an immobilized carbonic anhydrase (CA) core and allow water flow at right

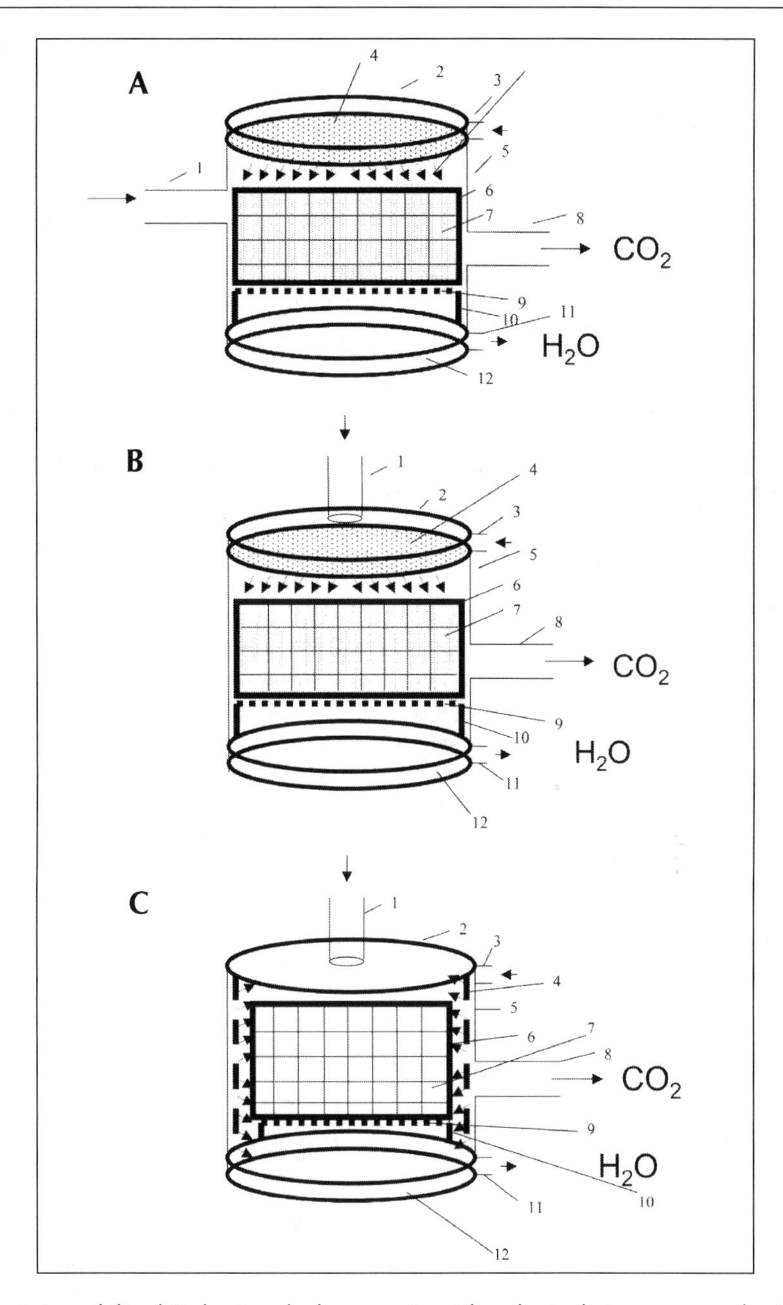

Figure 1. Immobilized Carbonic anhydrase reactors. Three basic designs were made: A) with horizontal gas flow and vertical water spray (flow); B) vertical inflow and horizontal outflow of gas and vertical water spray; and C) vertical inflow, horizontal outflow of gas (or vice versa) and horizontal water spray. The parts are: 1) inlet nozzle for gas/emission, 2) outer lid, 3) water inlet, 4) sprayer mesh, 5) the main vessel/reactor, 6) the large wire container for holding immobilized enzyme core, 7) immobilized carbonic anhydrase core, 8) outlet nozzle for gas/emission, 9) the bottom wire mesh for percolation of solution, 10) the holder stand, 11)water outlet, and 12) bottom solution holding chamber.

angles to gas flow, enabling solubilization of CO_2, which is then fed to coupled, immobilized Rubisco reactors. Capturing carbon dioxide from the emission stream has become possible by using immobilized carbonic anhydrase on a highly porous matrix core reactor where emission gas flows at right angles to water flow where CO_2 is solvated by the enzyme. The porous reactor core and water flow at right angles help significantly reduce pressure of the emission stream while at the same time allowing capture of CO_2. However, the capture step, in order to adapt for a continuous fixation scheme, needs continuous catalytic stability with respect to impurities and long operational life with minimum maintenance, including change of immobilized enzymes.[10,11] The porous core minimizes pressure drops and immobilized CA allows for efficient capture of CO_2. Thus, a biotechnological process where the carbon dioxide present in the emission stream (free of soot) could contact with water in the presence of immobilized carbonic anhydrase,[11] resulting in catalytic solubilization of carbon dioxide in water, has been developed. The contacting process of gas with water also results in the concentration of CO_2 from the emission stream in the aqueous phase. The factors that affect the CO_2 capture by these devices are gas and water flow rates, reactor core length to diameter ratio, average immobilization matrix pore size, core thickness and number of reactors with incremental volume that sums up for a single reactor of combined volume. The operation and stability of these reactors, method of immobilization of CA, and further details of process parameters on CO_2-capture efficiency can be found in published reports.[10,11] Advanced research along these lines is likely to enable enhancement in enzyme stability in the presence of a continuous flow of water, air and impurities. Capture of CO_2 from emission with high specificity and efficiency is necessary. Captured carbon dioxide in solution is suitable for fixation with coupled Rubisco reactors.

Bioreactor Operation and Stability of the Immobilized Biocatalyst

The CA reactor shown in Figure 1 houses the immobilized enzyme core. The purified recombinant carbonic anhydrase was used in the reactor core. The enzymes were immobilized on a glass-, polystyrene- or silica-coated steel matrix of different average mesh size using methods as reported earlier.[11] A thin film of water around the enzyme in the immobilized microenvironment keeps the enzyme hydrated and active for a long time; buffering of the enzyme apparently is not necessary for a long period. Buffer flushing of the reactors every 72 hours of continuous operation greatly enhances the shelf-life of the immobilized enzyme. Three basic designs of the reactors have been presented. The reactors in all three designs provide the ability to control two different flows: flow of emission gas and that of water spray. With respect to the flow of gases it is either horizontal inflow and horizontal outflow or vertical inflow and horizontal outflow (or vice versa). With respect to water spray, it is either vertical or horizontal. Therefore, the basic design of the reactor was reduced to three different types (a) with horizontal inflow and outflow of gas and vertical water spray, (b) vertical inflow, horizontal outflow of gas (or vice versa) and vertical water spray, and (c) vertical inflow, horizontal outflow of gas (or vice versa) and horizontal water spray (Fig. 1A-C).

Assay of Carbonic Anhydrase

Carbonic anhydrase activity is assayed using an electrometric method.[13] The assay is initiated by the addition of 10 ml of ice-cold, CO_2-saturated water into the reaction vessel. The change in pH from 8.0 to 7.0 at 25°C was monitored using a benchtop pH meter and semi-micro combination electrode with the signal directed to a chart recorder. CA activity is expressed in Wilbur-Anderson (WA) units per mg of protein and is calculated using the formula [(t0/t-1) x 10]/mg protein, where t0 and t represent the time required for the pH to change from 8.0 to 7.0 in a buffer control and CA sample respectively.

Immobilization

Carbonic anhydrase was immobilized using different coupling methods on a steel matrix coated with glass, polystyrene or silica.[11] The iron filings collected from a lathe machine were used for silanization with about 10 mg CA in Tris or HEPES buffer pH 8.0 per gram matrix. The inorganic support material is first treated with organo-functional silane.[14] The reaction of the carrier with gamma-aminopropyl-triethoxy-silane was used for coupling. Silane polymerizes across the surface of the carrier are anchored at intervals.[14,15] The amino derivative was covalently coupled using carbodiimides as described for other enzymes and matrices[16] or converted into carboxyl derivative using an alkylamine-carrier with succinic anhydride.[17] A very thin layer of glass was coated on iron filings and used for direct attachment of carbonic anhydrase using cyanogen bromide mediated coupling.[18,19]

Measurement of Pressure and Carbon Dioxide in the Emission Gas

The pressure of the emission stream in the inlet and outlet was measured using a HD8804 K kit or Testo 525 instrument; carbon dioxide in the emission gas was measured using Testo 400 IAQ kit equipped with an 0632 1240 and 0635 1240 CO_2 probe (Hotek Technologies, Tacoma WA).

Fixation of Carbon Dioxide

A CO_2 fixation technology utilizing immobilized Rubisco and 5-carbon D-Ribulose-1,5-bisphosphate (RuBP) results in formation of two three-carbon 3-phosphoglycerate.[7,8] Immobilization of Rubisco was performed and characterized for reactor operations [16]. Further advancement along these lines will enable continuous operation of the immobilized Rubisco under condition of continuous acceptor supply in the presence of impurities in the reaction mixture. The usefulness of this approach have been shown using immobilized mesophilic Rubisco.[16,20]

De Novo Synthesis of RuBP for Start-Up Operations

To the process of fixation, carbon dioxide from the emission stream and RuBP are necessary. RuBP regenerated from 3PGA is not suitable for start up operation. A separate scheme is required for generation of RuBP as shown in Figure 2. This scheme results in formation of RuBP.

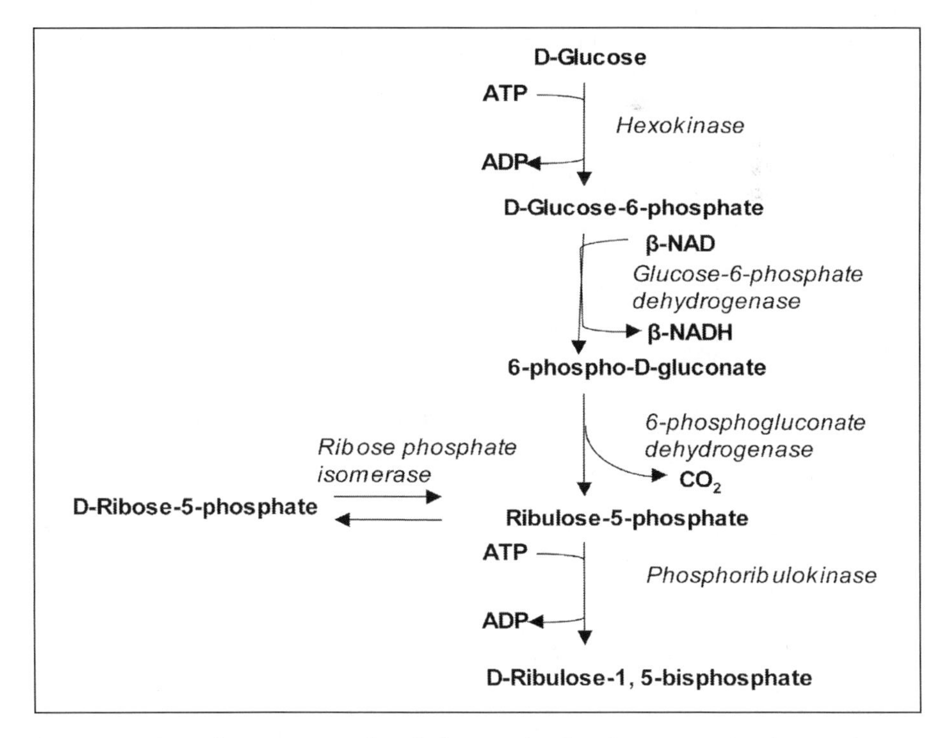

Figure 2. A scheme for preparation of D-ribulose-1,5-bisphosphate preparation from D-glucose. This scheme enables reactors to produce RuBP for start-up operations for fixation of CO_2 from emission sources.

This scheme is amenable to scaling and may generate RuBP at a large scale for start-up operations. The generation of RuBP requires concerted conversions by several immobilized enzymes as shown in the scheme (Fig. 2). This scheme is very different from RuBP regeneration from 3PGA (the product of CO_2 fixation) that is necessary for continuous CO_2 fixation.[21] Properties of the enzymes and their assay procedures are reported elsewhere.[21]

Fixation of CO_2 by Immobilized Rubisco

Rubisco purified from mesophilic sources, spinach, maize and wheat leaves,[22] were immobilized by 1,3-dicyclohexylcarbodiimide (DCC) or dimethylpimelimidate (DMP) coupling to the immobilization matrix. The immobilized Rubisco was found to be more stable than soluble enzyme at 55°C and showed stable activity for up to 50 days at room temperature. Operation variables and their effect on performance on immobilized were also determined.[16]

Regeneration of Acceptor for CO_2 Fixation

The successful continuous fixation of carbon dioxide depends on the rate of acceptor regeneration. As shown in Figure 3, and listed in Table 1, the linear combination of over-expressed recombinant enzymes were used in a cascade of reactors (11 reactors were employed in this scheme) for regeneration of RuBP from 3PGA generated as a result of CO_2 fixation.[8,23] The performances of the cascade of 11 reactors with respect to operation variables were determined.[23] Figure 4A shows the RuBP regeneration reactors using linear combination and Figure 4B shows the reactors as shown schematically in Figure 5. The regeneration process should occupy minimal weight and volume. Internal channeling within the enzymes acting sequentially on the starting material may significantly enhance the regeneration rate of acceptor. The significant reduction in volume and weight can be achieved by employing enzyme complexes or interactomes instead of the linear combination of reactors currently practiced.[20] Reduction in weight and volume would be useful from a maintenance and efficiency standpoints in stationary emitters, but is necessary for RuBP regenerating reactors in order to apply in mobile emission devices. Inside

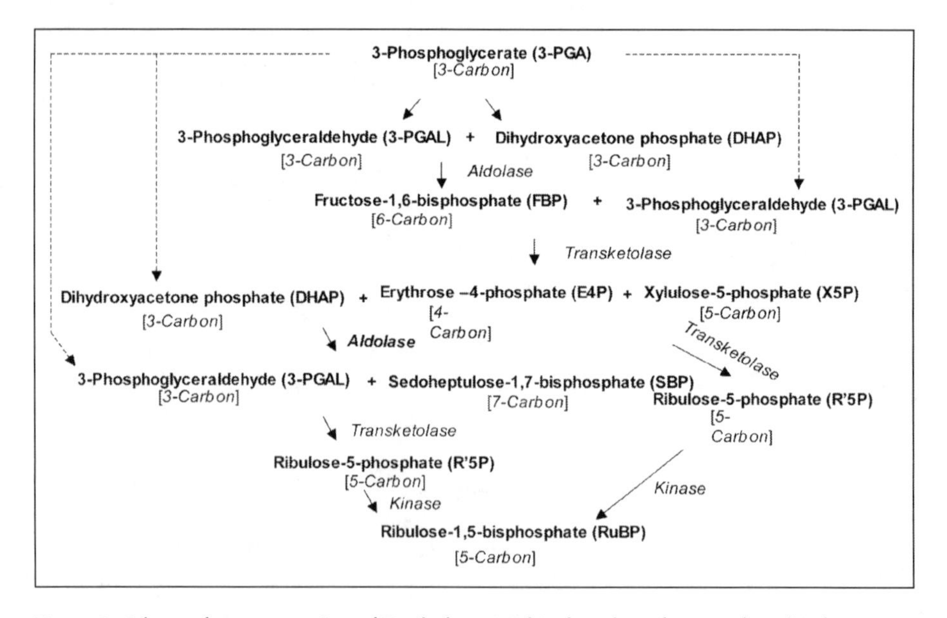

Figure 3. Scheme for regeneration of D-ribulose-1,5-bisphosphate from 3-phosphoglycerate.

Table 1. Immobilized enzymes used in RuBP regeneration ensemble of reactors

Enzyme	Km (mm)	Vmax (U/mg)	Reactor Length Designed (cm)	Reactor Length Constructed (cm)
Phosphoglycerate Kinase (EC 2.7.2.3) pH 6.5 @ 25°C	1000	274	2.93	3
Glyceraldehyde 3-Phosphate dehydrogenase (EC 1.2.1.12) pH 8.5 @ 25°C	720	21	47.39	48
Triose-phosphate Isomerase (EC 5.3.1.1) pH 7.6 @ 25°C	650 (G3P) 7010 (DHAP)	1900 7380	1	1
Aldolase (FBP) (EC 4.1.2.1.3) pH 7.4 @ 25°C	20	35	5.26	6
Fructose-6-Phosphatase (EC 3.1.3.11) pH 9.5 @ 25°C	110	1670	1	1
Transketolase (EC 2.2.1.1) pH 7.7 @ 25°C	205	10.5	35.57	36
Aldolase (SuBP) (EC 4.1.2.1.3) pH 7.7 @ 25°C	3810	98	27.44	28
Sedoheptulose 1,7-bisphosphatase (EC 3.2.3.11)	560	54000	1	1
Transketolase (EC 2.2.1.1) pH 7.7 @ 25°C	510	10.5	48.11	48
Ribulose 5P-3-epimerase (EC 5.1.3.1) pH 8.0 @ 25°C	200	11900	1	1.5
Ribulose 5P-kinase (EC 2.7.1.19) pH 8.0 @ 25°C	90 (Ru5P) 55 (MgATP)	380	1	1.5

the cytoplasm, a vast number of the protein work next to each other without apparent compartmentalization and yet they show a great degree of specificity. At the same time, within the same viscous environment embedded within many other components, many enzymes are able to convert their substrates often thousands of folds more rapidly in response to stimulus than that at base level. Such rapid and specific conversion is attributed to channeling. Knowledge of metabolic networks[24] has led to the concept of interactome.[20] The interactomes are proteins or enzymes that interact for functional purposes but not in a strong, stable way but in a transient and often in non-stoichiometric manner. Such interactions are keys to reducing volume and enhancing the enzymatic conversion rate. Although the knowledge of protein networks is emerging and it is becoming clear that proteins act in concert, creating artificial complexes where proteins will interact at least functionally and transiently is still a new thought pioneered by our team and we are still trying to develop new methods to overcome this obstacle. This will be an extremely difficult technological challenge.[20] The integration of interactomes or enzymatic complexes will enable efficient RuBP regeneration. Advances may result in the substantial reduction of number of reactors. This could be achieved, as mentioned before, by employing reactors that harbor enzyme complexes rather than a single purified enzyme. A detailed account of all enzymes and their assay and immobilization procedures is beyond the scope of this review and can be found in reports elsewhere.[7,8,23]

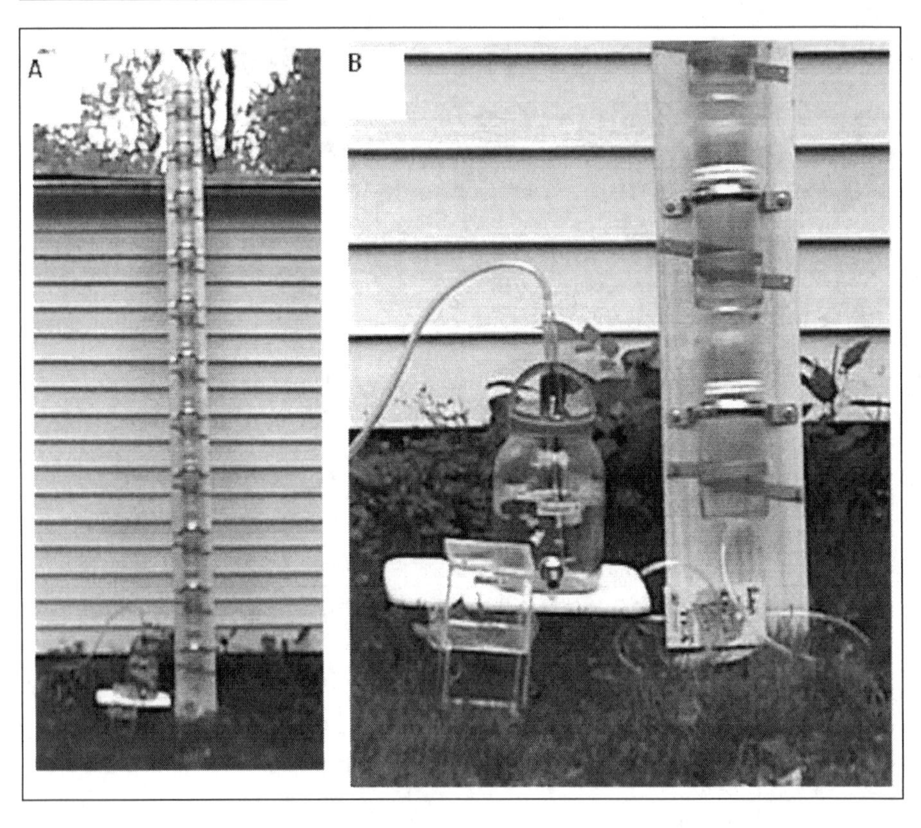

Figure 4. A) An experimental prototype device that was used for investigation of process variable optimization for RuBP regeneration (module C). B) The experimental Rubisco reactor (module A part 11), ATP generation system (module B) and RuBP regeneration device (module C) for capturing carbon dioxide from an experimental gasoline electric engine emission.

Cofactors for RuBP Regeneration (NADH and ATP)

Regeneration of RuBP, the 5-carbon acceptor for CO_2 fixation, needs cofactors NADH and ATP. A system to regenerate adenosine triphosphate (ATP) from ADP and inorganic phosphate using immobilized recombinant bacterial F0F1ATPase has been developed (Fig. 6A,B). The ATPase has F0 (subunits a, b and c) and F1 sectors (α, β and γ subunits). If the enzyme is immobilized via F0 sector on membrane it leads to ATP synthesis in response to potential gradient across membrane, but if another molecule is immobilized via F1 sector in response to same gradient it will hydrolyze ATP. The uniform immobilization thus is critical for ATP regeneration. A biotin-tagged or his-tagged subunit was used to specifically immobilize the F0 sector on membrane using recombinant enzyme from *E. coli*. Recombinant protein expression, purification and immobilization have been described in a report.[9,25] Continuous NADH regeneration has been mentioned in literature[26-28] and can be used for RuBP regeneration. An immobilized NADH dehydrogenase-based NADH regeneration system coupled with or without a specific NAD capture system will allow its use with RuBP regeneration reactor system. Recombinant NADH dehydrogenase derived from thermophiles may be purified in *E. coli* subjected to covalent attachment using different chemical methods. Uniform orientation, during immobilization, can be achieved using used methods for F0F1 ATPase.[9]

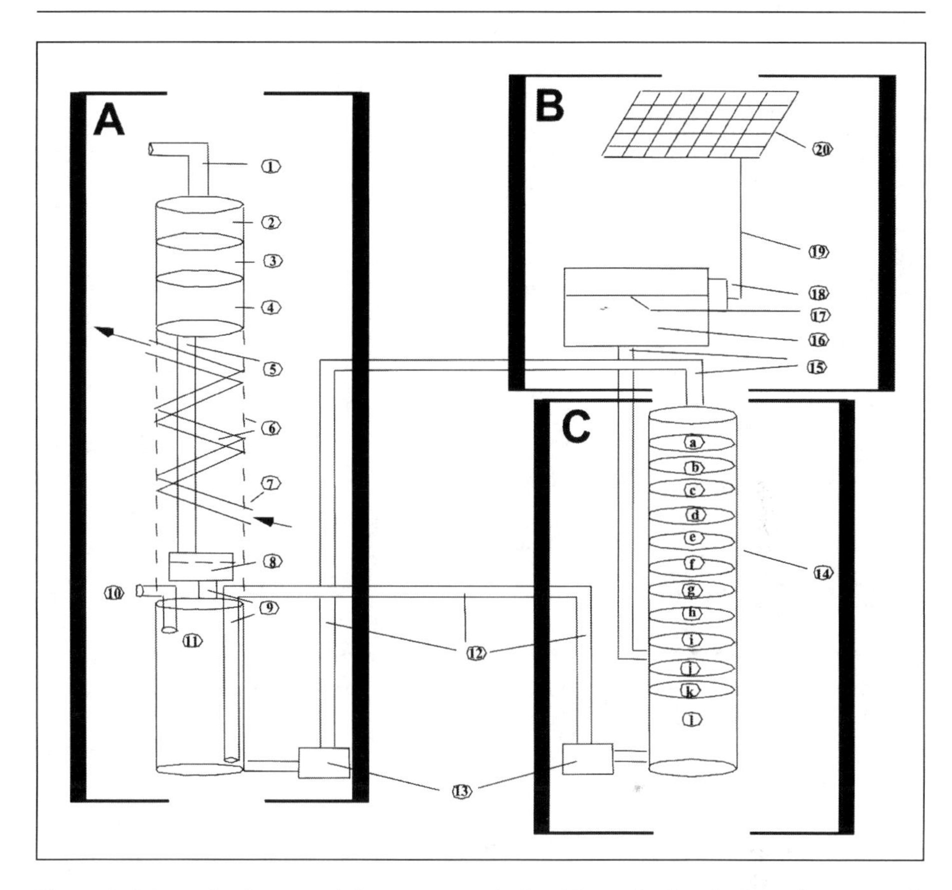

Figure 5. Schematic diagram of the prototype device (shown in Fig. 4). Parts shown are: 1) Connector to the internal combustion engine exhaust; 2) Catalytic reduction chamber; 3) Catalytic oxidation chamber; 4) Solid/Liquid SO_2 trap; 5) Connector tubing for exhaust gas; 6) Counter current heat exchanger; 8) Water reservoir for trapping SO_2; 9) Connector tubing; 10) Safety exhaust; 11) Chamber for catalytic conversion by Rubisco; 12) Connector tubing; 13) Pump; 14) An ensemble of multi plug flow reactors. A 13+2 series of reactors are identified by alphabets harboring different enzymes: a) Triose-phosphateisomerase, b) Aldolase, c) Fructose-6-phosphatase, d) Transketolase, e) Aldolase, f) Sedoheptulose-1,7-bisphosphatase, g) Transketolase, h) Ribulose-5 phosphate epimerase, i) Ribulose phosphate isomerase, j) Ribulose phosphate kinase; 15) connector tubing; 16) ATP generation chamber; 17) Semi-permeable membrane; 18) Converter and connector panel for electrochemical gradient; 19) Connector to wire solar panel; and 20) Solar panel and electrochemical gradient converter.

Economic Benefits of Carbon Dioxide Fixation

The detailed cost analyses of a previous experimental system that uses the scheme outlined in Figure 5 have been determined;[7,25] the cost is influenced by the energy consuming step and for a solar energy (8-12h), or electrical energy (12-16h) driven model the present bioprocess is very cost effective: for every 114g of octane about 613g of 3PGA is regenerated. This provides a cost advantage of about $30,000 per annum enabling recovery of an initial fixed cost of $55,000 for a bioprocess with capability of handling 0.12 L/h octane emission. This generates a substantial operating profit and leads to recovery of fixed cost investment within a period of

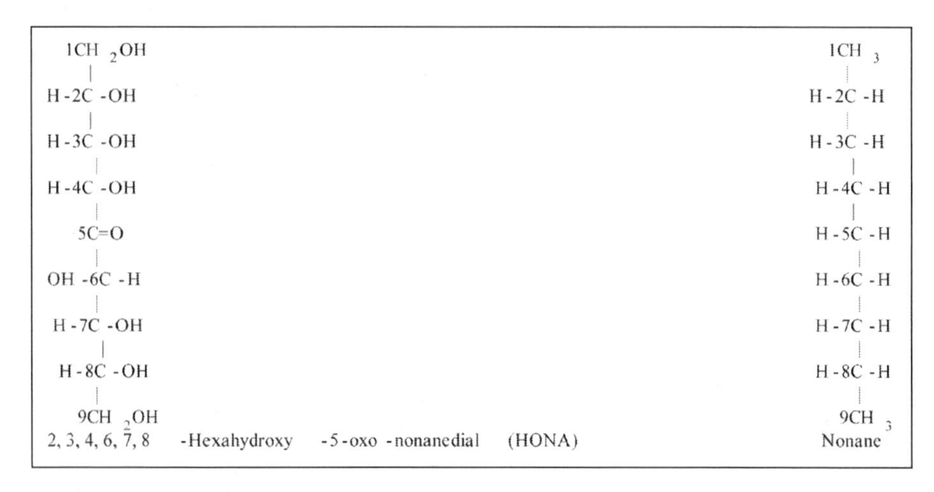

Figure 6. Representative (Fisher projection) formula of HONA (2, 3, 4, 6, 7, 8-hexahydroxy-3-oxo-nonanedial), the synthetic molecule that results from the combination of three 3PGA (3phosphoglycerate). The conversion of HONA into Nonane requires vigorous hydrogenation with several catalysts.

Table 2. The estimate of molar requirement of acceptor for CO_2 fimation

	Mole	Amount/Amount	Density	Rate	Rate
C_8H_{18}	1	114g	0.98	1liter/hr	16.3g/min
CO_2	8	352g	NA	3025g/hr	50.4g/min
RuBP	8	2480	NA	21364g/hr	356.1g/min

Table 3. The reactor requirements for CO_2 fixation/RuBP regeneration

	Fixation of CO_2	RuBP Regeneration
Solubility of RuBP	100g/L	200g/L
Rate of utilization (C_8H_{18})	0.12 L/h (114g/h)	5 L/h (4900g/h)
Reactor Voume (-regeneration)	25L	534L
Reactor Voume (+regeneration)[a]	1.3L	26.7L

[a]Considering rate of RuBP conversion and regeneration = 42 and 1780 g/min respectively, start-up amount of RuBP in the reactor with regeneration is 3XAverage Residence Time of RuBP

about 4-5 years. The cost may be driven down with mass production, and with the use of recombinant and relatively stable hyperthermophilic enzymes. The 3PGA is a concatenated carbon compound and can be easily converted into amino acids, fatty acids, antibiotics or nucleotides. In cost calculations, average saleable price of 3PGA or an easily convertible a product (Glycine) thereof was taken into consideration as byproduct. The economic analysis assumes octane as the fuel burned and the prototype device was performed based on rates and reactor parameters provided in Tables 2 and 3.

Potential for Hydrocarbon (Fuel) Generation

As shown in Figure 6, we have been able to generate a new 9-carbon concatenated from extracted 3PGA (byproduct of CO_2 fixation) named HONA and convert it into the 9-carbon hydrocarbon Nonane. The composition of the precursor of HONA was verified by TLC separation and elemental analysis. This was achieved by molecular ligation of three 3PGA molecules using process and catalysts developed by our group. The full economic value of these conversions (3PGA to Nonane) will be realized only when this approach is applied at a scale that may represent industrial emission.

Potential Conversion of Fixed Carbons into Other Chemicals

The 3PGA extracted from a pool of fixed carbon dioxide can be utilized for conversion into other organic compounds. A comprehensive account of these conversions is outside the scope of this review, but an example is conversion of 3PGA into amino acids. Reactors harboring permeabilized cells from keratinizing tissue such as skin and hair[29,30] in a bioreactor can be used for conversion of 3-phosphoglycerate into serine. Continuous production of serine, glycine and other amino acids can be made with 3-phosphoglycerate as the principal precursor material in such reactors employing keratinizing tissues.

Conclusions

This chapter presents the novel use of enzymes in a complex fashion to abate carbon dioxide pollution, while industrially generating a profit for polluting units. A scheme for continuous fixation of carbon dioxide from polluting industrial installations is presented. A set of four modules enable this conversion. The first module harbors carbonic anhydrase and enables capture of carbon dioxide from the emission stream. The second module harbors Rubisco-enabling fixation of carbon dioxide on RuBP generating 3PGA. The third module generates cofactors ATP and NADH and harbors enzymes F0F1 ATPase and NADH dehydrogenase respectively. The fourth module regenerates RuBP from 3PGA using a cohort of 11 enzymes. This is a complex system of several enzymes that helps a complex conversion of 3PGA into RuBP and thereby restarts the fixation step. A process harboring a cohort of enzymes and producing RuBP for start-up operation is also presented here.

References

1. Schnur R. The investment forecast. Nature 2002; 415(6871):483-484.
2. Joos F, Plattner GK, Stocker TF et al. Global warming and marine carbon cycle feedbacks on future atmospheric CO2. Science 1999; 284(5413):464-467.
3. Victor DG. Global warming: Strategies for cutting carbon. Nature 1998; 395:837-838.
4. Rahn T, Eiler JM, Boering KA et al. Extreme deuterium enrichment in stratospheric hydrogen and the global atmospheric budget of H2. Nature 2003; 424(6951):918-921.
5. Tromp TK, Shia RL, Allen M et al. Potential environmental impact of a hydrogen economy on the stratosphere. Science 2003; 300(5626):1740-1742.
6. Prather MJ. Atmospheric science. An environmental experiment with H2? Science 2003; 302(5645):581-582.
7. Bhattacharya S, Chakrabarti S, Bhattacharya SK. Bioprocess for recyclable CO2 fixation: a general description. In: Bhattacharya SK, Chakrabarti S, Mal TK, eds. Recent Research Adv in Biotechnol Bioengg. Trivandrum, Kerala: Research Signpost, 2002:109-120.
8. Bhattacharya SK. Conversion of Carbon dioxide from ICE exhausts by fixation. US patent 6258335, 2001.
9. Bhattacharya S, Schiavone M, Nayak A, Bhattacharya SK. Uniformly oriented bacterial F0F1-ATPase immobilized on a semi-permeable membrane: a step towards biotechnological energy transduction. Biotechnol Appl Biochem 2004; 39(Pt 3):293-301.
10. Bhattacharya S, Nayak A, Schiavone M, Bhattacharya SK. Solubilization and concentration of carbon dioxide: novel spray reactors with immobilized carbonic anhydrase. Biotechnol Bioeng 2004; 86(1):37-46.
11. Bhattacharya S, Schiavone M, Chakrabarti S, Bhattacharya SK. CO2 hydration by immobilized carbonic anhydrase. Biotechnol Appl Biochem 2003; 38(Pt 2):111-117.

12. Bhattacharya SK. Enzyme facilitated solubilization of carbon dioxide from emission streams in novel attachable reactors/devices. US patent application number 464789, 2003.
13. Wilbur KM, Anderson NG. Eletrometric and colorimetric determination. of carbonic anhydrase. J Biol Chem. 1948;176:147-154.
14. Bunting PS, Laidler KJ. Kinetic studies on solid-supported -galactosidase. Biochemistry 1972; 11(24):4477-4483.
15. Kobayashi T, Ven Dedem G, Moo-Young M. Oxygen transfer into mycelial pellets. Biotechnol Bioeng. Jan 1973;15(1):27-45.
16. Chakrabarti S, Bhattacharya S, Bhattacharya SK. Immobilization of D-ribulose-1,5-bisphosphate carboxylase/oxygenase: a step toward carbon dioxide fixation bioprocess. Biotechnol Bioeng 2003; 81(6):705-711.
17. Harhen B, Barry F. Immobilization of proteolytic enzymes. Biochem Soc Trans 1990; 18(2):314-315.
18. Srinivasan VR, Bumm MW. Letter: Isolation and immobilization of beta-D-glucosidase from alcaligenes faecalis. Biotechnol Bioeng 1974; 16(10):1413-1418.
19. Chikere AC, Galunsky B, Schunemann VV, Kasche VV. Stability of immobilized soybean lipoxygenases: influence of coupling conditions on the ionization state of the active site Fe. Enzyme Microb Technol 2001; 28(2-3):168-175.
20. Chakrabarti S, Bhattacharya S, Bhattacharya SK. Biochemical engineering: cues from cells. Trends Biotechnol 2003; 21(5):204-209.
21. Bhattacharya S, Nayak A, Gomes J, Bhattacharya SK. A continuous process for production of D-ribulose-1,5-bisphosphate from D-glucose. Biochemical Engg J 2004; 19:229-235.
22. Chakrabarti S, Bhattacharya S, Bhattacharya SK. A nonradioactive assay method for determination of enzymatic activity of D-ribulose-1,5-bisphosphate carboxylase/oxygenase (Rubisco). J Biochem Biophys Methods 2002; 52(3):179-187.
23. Bhattacharya S, Schiavone M, Gomes J, Bhattacharya SK. Cascade of bioreactors in series for conversion of 3-phospho-D-glycerate into D-ribulose-1,5-bisphosphate: kinetic parameters of enzymes and operation variables. J Biotechnol 2004; 111(2):203-217.
24. Bhattacharya S, Chakrabarti S, Nayak A, Bhattacharya SK. Metabolic networks of microbial systems. Microb Cell Fact 2003; 2(1):3.
25. Bhattacharya S, Schiavone M, Nayak A, Bhattacharya SK. Biotechnological storage and utilization of entrapped solar energy. Appl Biochem Biotechnol 2005; 120(3):159-168.
26. Egorov AM, Osipov AP, Pozharskii SB, Iavarkovskaia LL. Kinetics of the NADH regenerating system using bacterial formate dehydrogenase. Biokhimiia 1981; 46(2):361-367.
27. Egorov AM, Tishkov VI, Popov VO, Berezin IV. Study of the role of arginine residues in bacterial formate dehydrogenase. Biochim Biophys Acta 1981; 659(1):141-149.
28. Egorova OA, Avilova TV, Platonenkova LS, Egorov AM. Isolation and properties of NAD-dependent formate dehydrogenase from the yeast Candida methylica. Biokhimiia 1981; 46(6):1119-1126.
29. Goldsmith LA, O'Barr T. Serine biosynthesis in human hair follicles by the phosphorylated pathway: follicular 3-phosphoglycerate dehydrogenase. J Invest Dermatol 1976; 66(6):360-366.
30. Malgrange B, Rigo JM, Coucke P et al. Identification of factors that maintain mammalian outer hair cells in adult organ of Corti explants. Hear Res 2002; 170(1-2):48-58.

INDEX

K

L

M

N

P

R

S

T

W